Table of Contents

Learning Resource Centre

Email: LRC@swindon.ac.uk   Tel: (01793) 491591 ext.1500

CMP**Books**

San Francisco and New York

Stay informed: Want to receive email news updates for The Complete
E-Commerce Book? To subscribe and receive occasional updates,
changes, and news related to this book, please send a blank email to
ecommerce@news.cmpbooks.com

Published by CMP Books
An imprint of CMP Media LLC
Main office: CMP Books, 600 Harrison St., San Francisco, CA 94107 USA
Phone: 415-947-6615; Fax: 415-947-6015
www.cmpbooks.com
Email: books@cmp.com

CMP
United Business Media

ISBN: 1-57820-312-0

For individual orders, and for information on special discounts for quantity orders,
please contact:
CMP Books Distribution Center, 6600 Silacci Way, Gilroy, CA 95020
Tel: 1-800-500-6875 or 408-848-3854; Fax: 408-848-5784
Email: cmp@rushorder.com; Web: www.cmpbooks.com

Distributed to the book trade in the U.S. by:
Publishers Group West, 1700 Fourth Street, Berkeley, California 94710

Distributed in Canada by:
Jaguar Book Group, 100 Armstrong Avenue, Georgetown, Ontario M6K 3E7 Canada

Cover design by Robbie Alterio
Text design by Brad Greene, Greene Design

Transferred to digital printing in 2009.

Preface

The dot-com bubble has burst, but e-commerce isn't dead — online sales are still growing impressively. In the U.S., Forrest Research reported that 2002 online sales were up nearly 50 per cent compared with the year before. And in the U.K., the Interactive Media in Retail Group said that 2002 online shopping grew 19 times faster than traditional retailing venues. Furthermore, as of December 17, 2003, The Street.com reports "e-commerce sales are up 20% to 35% this holiday season [2003]." In fact, numerous experts predict that Internet shopping is likely to continue to outgrow brick-and-mortar sales for the foreseeable future because consumers are recognizing and exploiting the benefits of Web shopping.

A late 2003 eSpending report found that 62 percent of the online shoppers interviewed described themselves as satisfied with the experience, while less than 7 percent were dissatisfied. This means that online shoppers are also more confident about security, and thus, more comfortable spending money online.

2003 was the year that e-commerce proved it is not a "flash in the pan." The Web is now a common shopping venue for the average consumer.

You are reading this book because you want to get in on this shopping bonanza. However, building and running a successful website isn't as easy as it might first appear. Like the first edition of *The Complete E-commerce Book,* this edition provides select comparisons of various tools needed to construct, design, and run a successful e-commerce website. New to this second edition are sections on auctions, peer-to-peer file sharing, and weblogs. As such, this edition provides details on how to build an online auction site or incorporate auction features into an existing site. The same holds true for peer-to-peer file sharing and weblog technology. Furthermore, the reader is given suggestions on how to turn a file sharing venture or a weblog into a moneymaking venture, whether as a new e-commerce venture or as an add-on to an existing website. Adding an auction, peer-to-peer file sharing, and/or a weblog to an established e-commerce business are great ways to grow a community, create repeat customers, and build a loyal consumer base.

The technology that makes the Web a wonderful place to do business also makes it scary. Why? Think about it — the Web is unstructured, uncensored for the most part, and it speeds up the exchange of information between businesses and their customers — at a speed that, if you are not careful, you will not be prepared to handle. Your customers can find your business by a click of a mouse and make purchases easily but they also expect easy access, prompt delivery, and good customer service. This book gives you the tools to understand the technology and issues involved; thus, enabling a business to adapt quickly and to make the changes necessary to prosper on the Web.

The tools that are now available enable e-commerce operators to create more exciting, dynamic websites. This second edition addresses these new products. There is a new chapter, "Search Engines and Directories," where you will find a lengthy discussion on how to design your website to receive a high "ranking" within the various search engines and directories.

This second edition also delves further into security issues. The Internet is a playground for hackers and other nefarious individuals. If an e-commerce operator isn't careful, his or her website can become a vehicle for fraud. But, like in the real world, there are steps you can take to keep out the "bad guys." This book is peppered with new advice on how to protect your website, your servers, and your data.

For those readers who already have an established brick-and-mortar business, an online presence helps to give them a competitive edge. Understanding the basics of e-commerce and its technology is necessary for any executive in a company that is already on the Web, thinking about moving to the Web, or has business partners who have moved to the Web. For a successful web presence you need not only a well-designed and easily maintained website, you also must bring together product knowledge, marketing, distribution and other skills that may not be present in a brick-and-mortar business. *The Complete E-commerce Book* helps you to ask the right questions and set realistic goals, and to make the right decisions as you move to the Web.

I wrote this book with both the entrepreneur and the non-technology executive in mind. Both can, at times, find e-commerce and its obscure technological terms confusing. When it comes to making the decision to adopt new technologies, an executive relies on his or her information technology (IT) personnel. However, executives who rely exclusively on these technical gurus for significant input into their business' overall strategy are just asking for trouble. Why? While non-technology executives should consult with the IT managers on how technology can relate to their business' overall strategy, executives must take responsibility for understanding the evolving technologies so that they can make the best decisions possible to ensure the success of their e-commerce venture.

Janice Reynolds
Oley, Pennsylvania
February 5, 2004

Chapter 1

The E-Commerce Phenomenon

It is hard to fail, but it is worse never to have tried to succeed.
In this life, we get nothing save by effort.

—Theodore Roosevelt

Despite the spectacular dot-com bust of a few years ago, the Internet has markedly changed the way we do business, whether it's finding new streams of revenue, acquiring new customers, or managing a business's supply chain. E-commerce is mainstream — enabling businesses to sell products and services to consumers on a global basis. As such, e-commerce is the platform upon which new methods to sell and to distribute innovative products and services electronically are tested.

The Web's influence on the world's economy is truly astonishing. The business world knows that the Web is one of the best ways for business such as manufacturers to sell their products directly to the public, brick-and-mortar retailers to expand their stores into unlimited geographical locations, and for entrepreneurs to establish a new business inexpensively.

Thus, it is important that the executive in the 21st Century know 1) where technology stands in the business processes of his or her company, 2) how technology relates to the company's strategies, 3) how rapidly technology changes and evolves, and 4) how the company and its business partners will respond to the changing technology.

In the high flying 1990s, many people jumped on the e-commerce bandwagon after reading the many highly publicized dot-com "success" stories. Admittedly, most were written to raise the entrepreneurial blood pressure. What many forgot, though, was the old adage: If it looks too good to be true, it probably is. They didn't use their innate intelligence and failed to proceed with caution.

Nonetheless, the ascendancy of e-commerce has expanded the business environment so that even a small start-up can compete with well-established business names and product brands. Yet, when you consider joining the e-commerce commerce community, keep in mind that selling products and services on the Web presents a unique set of challenges. This book will help you take on those challenges.

Note: *U.S. Department of Commerce reports show that U.S. online retail sales totaled $14.33 billion during the fourth quarter of 2002. The reports further show that this amount represents an increase of 28.2% over the same period in 2001.*

✪ THE E-COMMERCE DECISION

For both the entrepreneur and the established brick-and-mortar business, setting up a virtual business on the Web can be a great way to expand your market — BUT you MUST first educate yourself as to how to proceed. Extending an existing business to the Web or starting a new Web venture without understanding e-commerce is like trying to hit a baseball with one arm tied behind your back and a blindfold on. It is not simple and it is not easy to build a successful business on the Web. It takes a tremendous amount of planning. Without a clear understanding of the process and clearly defined goals, establishing a successful website can turn into a difficult, complicated, and costly project.

Back during the dot-com craze, the author read a Tom Friedman column in *The New York Times* entitled "Amazon.You: For about the cost of one share of Amazon.com (AMZN), you can be Amazon.com." In that column Friedman told the story of a small Midwest online book-selling operation that cost around $150 a month to run. That small bookseller used the same wholesalers as the competition, including Amazon.com, but his net on each sale was better. Friedman stated that the site "offers Millions of Books at Great Prices" including, a John Grisham book for $2.60 less than Amazon.com. Mr. Friedman's point was that for the cost of just a few shares of Amazon.com stock, and a *good, clearly developed idea*, a savvy small entrepreneur could compete with a "Web Giant."

The Brick-and-Mortar's Decision

Are you the owner of an established, traditional business considering a move to the Web? If so, before moving to the Web, you must first determine the benefits of a Web presence. To do so, ask yourself some hard questions:

* What are your business goals and how will a website support them?
* What do you want to achieve with a website?
* Will a website deliver new customers or more sales from current customers?
* Will it offer better customer support?
* Will it bring good publicity?
* When is the right time to create a website for your business?
* Will this venture adversely affect your brick-and-mortar business?

In other words, if you build it will they come? Forget the hype. Determine if and how a website can enhance your business's bottom line. Then determine the amount of money you must invest before you turn your first profit from the website.

Any way you look at it, the decision to extend an already successful business onto the Web is a tough call. Before moving forward, considerable research is necessary; without the appropriate technical knowledge and the understanding of the ins-and-outs of e-commerce, it is easy to get lost.

On the technical side, your first sequence of goals should be:

* Learn what it takes to build various kinds of websites.
* Determine what kind of website you want to build.
* Build it.
* Continue to improve and update it.

You must take care that your existing business doesn't suffer as you and your staff gain the necessary knowledge and proficiency in the technologies needed to purchase and to implement the equipment and software needed for a website. Do you have any idea how much time, effort, and money you must invest in order to have a viable e-commerce site, one that will produce a reasonable return? Take the time to find out.

An Entrepreneur's New Web Venture

Most of what applies to a brick-and-mortar business also applies to an entrepreneur's new web venture. But a new e-commerce business also must strive to leave the starting gate with an extremely high performance level if it is to establish its brand name effectively. If it stumbles, in all likelihood that new web-based business will not positively imprint its brand name on the public and therefore will not be in a position to exploit the opportunities available to an innovative web entrepreneur.

For an entrepreneur, the secret to a successful web-based venture is finding a niche. The first step for a fledgling entrepreneur is to determine your niche market — a defined group of potential customers sharing common characteristics that delineate their interest in specific products and/or services. A niche market makes it easier to plan a credible marketing strategy. When you know your niche market, you can tailor your site's content to appeal to that market. Examples of websites built to serve a unique niche market include 4mdogbooks, platesusa.com, and Gary's Belt Buckles, which can be found at www.abn1.net/buckles.

Once you decide on your niche market, and your product or service, ask the following questions:

* Is your product easy and economical to ship?
* Who are your suppliers? Can you depend on them for quality products and for prompt delivery?
* What is your price point (the price just above the point where lowering the price will increase your sales but not increase profits)?
* How will you draw customers to your website?
* How will you differentiate yourself from the competition?
* How will you accept orders and process payments?
* What type of fulfillment facilities will you use to ship products?
* How will you handle returns and warranty claims?
* How will you provide customer service?
* What size e-commerce site would you want to build?
* How will you build your new website?
* How much will it cost to build the website, and can you afford it?

✪ THE ESSENTIALS

You can't just open an online store and expect customers to flock to it. Find out if your niche market is one that you can reach through a website. How? Does your niche market have an identifiable need for your web offering? Do they have the wherewithal to pay for it? Is the niche group sizeable, i.e., will it provide enough business to produce the income you need? If the answers are yes, you have found a good niche. Now dig deep within that niche to understand the consumer behaviors that drive it.

Every e-commerce operator should assume that his or her customers are sophisticated shoppers who demand prompt delivery of a product that is exactly as portrayed on their website. The most common mistake made by inexperienced web operators is to fail to be responsive to their customers' order processing and fulfillment needs. But those services are the very underpinnings of all successful e-commerce ventures — neglect those areas and you have a business catastrophe.

To help in the follow-through, you and your customers must be able to track the status of each purchase. Most new e-commerce businesses, however, fail to integrate this necessary backend support.

Another "must" is to make certain that your customers know that your web-based business *will not only deliver* a value online that cannot be found offline, *but* that it is just as responsive with customer service issues as the most well-regarded offline business.

By keeping customer service and product fulfillment as an immediate priority you can build a valuable relationship with your customer. In doing so, you earn that customer's loyalty. That helps to stem the natural flow of attrition as customers who pursue the lowest price find that the trade-off is a void in the cut-rate business's customer service department.

Another common problem for new e-commerce businesses is misinterpreting the power of the Web. Yes, a website with the right infrastructure can economically automate transactions. However, the real power of the Web is its role as a relationship-building magnet — through its ability to provide numerous opportunities for interactivity. If you are careless with automated processes — this very real advantage will vanish.

Use your website to provide not only useful and interesting information about your products/services, but also about your entire niche market. The group that makes up a niche market always yearns for more information. They will return time and again to your website if they are appealed to on the basis of their special interests — detailed articles and content-rich advertising specifically targeted to them.

The dot-com bust of 2000 was a failure of business plans; the concept itself has not failed. And while numerous news articles over the last few years detail how various websites lost sales and customer confidence due to inadequate prelaunch planning, there have also been many successes, especially in the small business arena.

✪ WHY GO TO THE WEB?

The Web opens up a whole new market for goods and services. It creates opportunity for a multifaceted arena that offers new efficiencies for sales, marketing, customer service, shipment tracking, inventory monitoring, and many other aspects of the total business model.

Choice has always been the Holy Grail for consumers. Today's consumers have a wide variety of commerce choices: traditional businesses, mega discount stores, catalogs or direct market mail, and the Web. The Web, taken as a whole, is a powerful medium where consumers browse, research, compare, and then buy online or, after doing their "window shopping" online, make the purchase at a brick-and-mortar business. Businesses that keep in mind the consumers' desire for choice, and integrate into their website the appropriate means for customer interactions, will succeed.

This being said, the Web will not open vast new markets for every business. However, it can extend a significant degree of power to businesses that recognize how to leverage the efficiencies of this new arena. A good example is 3Com (www.3com.com) which, through its website:

* Offers products and services to a global audience.
* Provides many different technical support features online.
* Offers software downloads including drivers, updates and fixes, which prior to the website would have been mailed to the customer.
* Offers an online store.
* Provides an educational center with online courses.

In short, 3Com's business and customer base didn't change — the Web changed the way 3Com services its market — it did not create a new market. Still, overall, 3Com's business is enhanced by its web presence.

Big companies with plenty of technical expertise and buckets of money have always been able to build their own e-commerce systems, complete with a secure server, a high-speed Internet connection, and custom software. Luckily, costly e-commerce barriers are rapidly tumbling allowing any business to have a credible web presence. What was once expensive and difficult is quickly becoming affordable and easy to use.

Look Before You Leap

Established businesses shouldn't jump rashly onto the e-commerce bandwagon. They must first institute a well thought out plan upon which to build their e-commerce business model. If you neglect this important step, you risk losing your already established identity, your good reputation, and your customers. Two examples of not looking before leaping are the widely reported Toysrus.com 1999 holiday shopping season fiasco and the KbKids' website problems. According to survey conducted by Robertson Stephens, a leader in Equity Research, Investment Banking and Sales & Trading, in early 2000, 44% of the shoppers surveyed stated that due to their unhappy experiences with KBKids.com, it was unlikely that they would shop on the online toy retailer again. Perhaps if these retailers had scaled down their initial online effort as Bloomingdales.com did, their website and reputation would have fared better.

Both websites have learned from their mistakes. KBkids.com completely redesigned its web operation. But before creating the new website the KBkids.com designers, web producers, and writers conducted surveys and usability studies, participated in conference calls and held one-on-one meetings with KBkids.com shoppers. The result is a new sleek, innovative website with many customer-friendly features along with the functionality and back-office features needed to keep customers happy.

Toysrus.com took another tack — partnering with the etailing giant, Amazon.com. The two companies entered into a ten-year deal that draws on each company's core strengths. Amazon.com fulfills orders, handles customer service, and uses its expertise in front-end site design to build a powerful customer-support environment. Toysrus.com and parent company Toys 'R' Us identifies, purchases, and manages inventory, using its clout to get the hottest toy lineup. Revenues are split, and risks are shared.

The Challenge

Creating a business model for e-commerce starts with the following basic challenge: Can you define your company? Next, can you state your goals for the company? Finally, — within the aforementioned context — can you state what role e-commerce can play in helping your company maintain or change its identity?

The Top 10 Website Categories as of September 2003

Category	Unique Visitors as of August 2003	Unique Visitors as of September 2003	% Change
Retail-Food	8,324,000	9,732,000	16.9%
Weather	38,100,000	41,757,000	9.6%
Sports	48,169,000	52,633,000	9.3%
Retail-Flowers/Gifts/Greetings	23,576,000	25,373,000	7.6%
Entertainment-TV	44,085,000	47,154,000	7.0%
Hobbies-Lifestyles-Food	22,959,000	24,322,000	5.9%
Hobbies/Lifestyles-Home	26,384,000	27,909,000	5.8%
Online Trading	10,739,000	11,256,000	4.8%
Entertainment-Music	62,210,000	65,116,000	4.7%

Table 1. As this table indicates, online sales are growing steadily along with the total number of Internet users (148,811,000 in August 2003 to 150,045,000 users in September 2003). *Source: comScore Media Metrix.*

✪ WEBSITE MODELS

There are eight basic website models — ranging from the simple static pages of a brochureware site to richly interactive online gaming sites, to online stores chock full of products, to online auctions. Many websites combine several of these basic models. However, each model has unique characteristics that distinguish it from the other models and it is important to understand these differences.

Brochureware Site

A brochureware site is a marketing site that electronically aids in the buying and selling process. A traditional business often will build and maintain a brochureware site as a marketing tool with the objective of promoting the business and its products/services. A brochureware site is sometimes an adjunct to a business' technical support division providing online documentation, software downloads and a Frequently Asked Question (FAQ) section. Such a website can provide detailed information about the business's products/services, contact information including the business's address, telephone numbers, and email addresses. It can also be a tool to provide the public copies of a company's annual reports, press releases, and employment opportunities. Revenue from this kind of site is generated indirectly by creating an awareness of the business' products/services. All transactions occur offline.

Savvion develops business process management software that improves business performance and reduces costs within and across functional business units. Its website, www.savvion.com, is a good example of a high-quality brochureware site. It is clean, fast loading, and has all of the elements of a good website. To demonstrate the variety of businesses that take advantage of the Web to expand their business opportunities visit the following brochureware websites: Rolledsteel.com, Cohenhighley.com, Hayproperty.com.au, and Paulcato.com.

As you can see, you don't need to be a corporate giant to benefit from a brochureware site. In addition to the websites listed in the previous paragraph, here are two other examples of small businesses that use the Web to their advantage. First, is a local Chevrolet dealership in Reno, Nevada that has made the most from its brochureware site. Visit Championchev.com and you will find that the website is positioned to provide the sales and repair departments with additional sales, while reducing the business's overhead by eliminating many of the telephone calls requesting directions to the business, hours of operation, etc. The second is a Tennessee pharmacy that dispenses information and speeds prescription refills via their website. To check out the innovative ways this small business uses the Web to its and its customers' advantage, visit Wilsonpharmacy.com.

Online Store

An online store is a website where consumers buy products or services. This type of site is most commonly referred to as an e-commerce site or a "B2C" (Business to Consumer) site. In addition to most, if not all, of the content found in a brochureware site, an online store displays products/services along with detailed information (e.g. specifications and pricing) usually from a database with search features, and a method for online purchase. An online store must also provide extensive information about the products/services offered that not only aids in attracting consumers, but gives them enough confidence in the seller and the products/services to take the next step — making an online purchase.

One question the author is often asked is "what should an e-commerce site offer — online order processing, just a toll free number, or both?" The answer is: Offer both.

If you choose to take online payments, you must provide a secure, reliable, cost-effective system for authorizing payment and managing transactions. The best systems are based on the Secure Socket Layer (SSL) and/or Secure Electronic Transactions (SET) encryption technology, which provide the encryption of data and generate and display a "results page" to the customer following the transaction.

Further, a successful online store must be designed with the ability to store orders in a database or as tab-delimited text files so the data can be imported into an invoicing system. Then the website must be able to intelligently route encrypted email to the order fulfillment division.

A good example of a large online store is Healthtex.com. This is a great site in every respect. To see how a small brick-and-mortar business uses a web presence to enhance its bottom line, visit Parkaveliquor.com, where you will find an attractive, well-designed and fully functional e-commerce site that benefits both the storeowner and the customer. Other good online stores you might want to use as guides when designing your website include the Treliske Organic (www.nzsouth.co.nz/treliske), which offers "Certified Organic" Wool, Knitwear, Beef and Lamb; Badcataviation.com, a great toy airplane store; and Soccer Books Limited (www.soccer-books.co.uk) where you can find a huge selection of books, video and DVDs relating to the sport of soccer.

Another avenue that an entrepreneur might want to consider is to team with a large e-commerce site such as Amazon.com or e-bay.com to provide the e-commerce end while the entrepreneur provides the site's "content." Good examples of such a site are www.dolls-for-sale.com and www.politinfo.com.

Subscription Site

A subscription site targets a specific niche market that places a value on expert information, service, or a digital product delivered in a timely manner. Technical newsletters, access to research information, and graphics, music and computer game downloads are all examples of products and services that can be sold for a monthly fee, an annual subscription, or a small per transaction fee. While the revenue from such services and products should be able to fund the operating costs of a subscription site, in most instances, the income may not be substantial — the exception are sites that offer video, music, and/or computer game downloads.

A subscription site can process payments offline and provide via email a user's name and password for access, or it can provide a secure, reliable, cost-effective online system for authorizing payment and managing transactions. Again, the best systems are based on Secure Socket Layer (SSL) and/or Secure Electronic Transactions (SET) technology to encrypt the data, and generate and display a "results page" to the customer following the transaction.

E-commerce technology continues to become more sophisticated, and with every advance the financial prognosis for a subscription site should improve. A good example of a subscription site is iEntertainment Network (www.iencentral.com). This worldwide game and entertainment site offers both free ad-supported and fee-based online game channels. The site also offers a variety of monthly subscription plans.

The website Content-wire.com offers another example of a good subscription site that uses both banner ads and subscriptions for its revenue stream. This website offers a niche editorial product that covers a narrow vertical technology sector — web-based commerce. As such, it provides up-to-date news articles, produces features on a variety of subjects, and offers a good bit of research material. Although this website could be better designed, it is functional and the average surfer will find it easy to use. Because the site does derive some of its revenue from banner ads, some information offered is free (use it!). Note that surfers who pony up $100 or $200 for a subscription, there is much more data available for your perusal.

A sub-model that increasingly is finding favor is the website that offers downloadable content — via a subscription account, a la carte basis, or a combination of both. The most popular website within this sub-model is the new Napster.com. Although this digital music destination bears the same name as the famous, but defunct, peer-to-peer file sharing website, that business model is in its past. The current Napster.com is a digital music catalog site that also offers many rich community features for its customers. The new Napster.com allows consumers to choose how they want to experience music, offering both an a la carte store and a premium subscription service.

Not all digital music sites follow the subscription model, however. Apple's iTunes.com website, which at this writing offers more than 400,000 songs from a wide variety of musicians for less than a dollar per song, states that its iTune website offers music without the need to "agree" to complicated rules. There are no clubs to join, and no monthly fees — if you like a song, you just buy a downloadable copy of it for 99¢.

Check these sites out; they might give you some ideas on how you can make money from your own website.

Advertising Site

An advertising site is a content laden site, whose revenue base is the dollar amount derived from banners, sponsorships, ads, and other advertising methods. The traffic the site draws is the measure of its value. Recognized rating firms measure its value and then advertising rates are based upon that value. Important note: Very few sites can be supported entirely through advertising dollars.

Two good advertising sites are Cnet.com and Howstuffworks.com. Both are wonderful content laden advertising websites that every reader should bookmark. You should especially read the Howstuffworks website's explanation of how a web server works, which can be found at http://computer.howstuffworks.com/web-server.htm.

Another great advertising site is Thekidzpage.com. Although this colorful site's use of "pop-up" ads can be irritating, those ads allow this site to provide all of its great content for free. Children love to visit this site to play games and parents appreciate the more than 250 printable coloring books and learning activities.

But an advertising site needn't be graphic intensive nor host a variety of pop-up ads to be successful. Cases in point are the low-tech websites HomePCnetwork.com and Hrmguide.co.uk. Although these text-driven sites don't offer a lot of "bells and whistles," they make up for their lack of "eye-appeal" by their informative, easy-to-use content.

Online or Cyber Mall

A simple and easy way to sell products/services online is to open a shop in one of the many cyber malls on the Web. Online malls generally offer turnkey solutions for store creation, payment processing, and site management. For example, most cyber malls offer a template for implementing a catalog of products, a shopping cart application, and a form generator — allowing small businesses to quickly set up shop on the Web. The templated applications employed to set up a business's catalog and the ease of payment processing are tempting to a novice. These cyber malls also provide a high level of "click traffic." If you are new to the Web and would prefer to first "dip your toe in the water," a cyber mall may be the answer.

A cyber mall's biggest promise is to deliver more traffic to your "front door" than you would be able to do if you go it alone. Since you will be relying on the cyber mall's marketing savvy, make sure you verify that it can deliver. But note that the only "front door" which is advertised is the cyber mall's address, not your online store's address.

Most cyber malls offer a purchasing system, which enables a web start-up to avoid up-front shopping cart software costs, but a fee is accessed on each purchase — that fee will eventually exceed the cost of the software.

You need to ensure that the cyber mall you choose is one in which your online store can flourish. Consider the pros and cons of establishing your online store in a cyber mall including the restrictions and costs that some cyber malls impose. Then consider the option of setting up your own purchasing system and independent identity. For, in reality, a cyber mall is just a list of links categorized by store and product type.

Any number of cyber malls can help you get your website up and running in record time. Yahoo! Store, probably the most popular cyber mall on the Web, includes an easy tool that lets you register your own domain name (http://www.ourname.com) or, if you already have a domain name, they will help you transfer it. You are also given the option of using stores.yahoo.com/yourname, which does not require an up front registration fee. Freemerchants.com offers an easy and inexpensive way to set up an e-commerce site. Of course, like many cyber malls, with Freemerchants.com you are required to design your new website using Freemerchants.com's own online Store Builder (to which you can add your own graphics and backgrounds) — you upload sites or pages created in other programs to the freemerchants.com servers.

Or perhaps you would like to avail yourself of the services of a cyber mall that caters to a niche market. Regional cyber malls are one example of this type of online service. Danapointharbor.com promotes not only local online shops, but also other local industries and activities in the Dana Point Harbor, California area. And Outerbanks.com does the same for the North Carolina Outer Banks region. There are many of these regional services; if you find the concept interesting, check out what's available in your area.

Perhaps you want to present your product to a more defined niche, like the Indian artifacts mall at www.arrowheads.com/main.htm or TIAS.com's antiques cyber mall. Such an arrangement might be just the ticket to introduce your new e-commerce business to your targeted niche market.

Business-to-Business Site

All of the website models discussed in this book are built to serve either the individual consumer or the consumer and business customers. But a business-to-business site is built to serve other businesses; if the business also wants to serve its individual consumers, it usually builds a separate retail site to serve those customers. This book doesn't deal with the minutiae of B2B sites, since B2B sites, while using some of the technology discussed in this book, often also need other technology to further their goals of providing products/services to other businesses rather than individual consumers.

The growth of B2B e-commerce is explosive. For some businesses, B2B e-commerce already influences value chains, distribution channels, customer service, and pricing strategies. Others look to B2B for ways to leverage this new technology to increase sales, profits, customer loyalty, and brand preference. For detailed information on creating a viable B2B business model read books such as, *B2B: How to Build a Profitable E-Commerce Strategy,* by Michael J. Cunningham; *The eMarketplace: Strategies for Success in B2B eCommerce,* by Warren Raisch; and *B2B Application Integration: e-Business-Enable Your Enterprise,* by David S. Linthicum.

Auction Site

Auctions have been around for thousands of years. Traditionally, a person offers an item for sale and potential buyers bid on the item. The bidder willing to pay the highest price for the item wins the bid and takes the item home — the same with online auctions. The main difference between traditional auctions and web-based auctions is that the actual bidding and selling takes place over the Internet with interested buyers submitting bids electronically. The person with the highest bid at the timed close of the auction wins the bid and arranges to receive the item. The auction site acts as the middleman in the buying and selling transaction process.

There are a number of ways you can use the auction model. First, you could build an auction website and let that be your business model. Or you could add an auction component to another e-commerce model. Many auction sites are built or sponsored by major vendors who have an established website, and use their auction site either to attract customers, or to offer merchandise that is surplus, outdated, and/or seconds.

Many readers, however, will find established online auctions such as those offered by eBay, Ubid, Amazon auctions, and Yahoo auctions a way to build a credible e-commerce business. Using an established auction website to build a web-based business is inexpensive and allows you to begin making a profit immediately: There are none of the expenses of the typical e-commerce model — no advertising costs, no hosting costs, etc. Auction sites receive billions (yes billions) of visits daily.

If you use one of the online auction sites as a means to enter the world of e-commerce, understand that you are responsible for listing your items on the auction site, and you assume responsibility for all aspects of your auction listings, including product descriptions, identification of quantities, establishment of starting and maximum bid prices, and shipping. Once the auction closes, it is up to you and the buyer to make arrangements for payment and shipping.

As online auctions continue to grow in popularity, more and more, entrepreneurs, retailers, manufacturers, and other businesses see them as a beneficial way to sell surplus goods, while consumers see them as a great way to save money and get great deals.

The largest and best known auction site is eBay.com. For sellers, the ability to market your product to millions of daily visitors makes using eBay one of the most efficient ways to sell just about anything. Potential buyers search for items and place bids on those they are interested in purchasing. At the close of an auction the highest bidder is the winner. At that point, the buyer and seller make arrangements for payment and shipping.

All you need to become an eBay seller is to register, which is free and only takes a few moments. However, eBay requires sellers to provide a credit or debit card, as well as enough details for eBay to check their "bonafides," i.e. the potential seller provides eBay with enough information for the auction site to establish his or her proof of identity by cross-checking against consumer and business databases. If everything is kosher, the seller obtains an "ID Verified" designation, which enables them to place items for sale on the auction site.

Yahoo! also offers an auction service. Its seller requirements are similar to eBay's, i.e. there is a registration (free) and all sellers must provide credit or debit card information.

If you have products that are more niche oriented, you might want to use an auction site that caters to that niche crowd, e.g. Allbadges.com is a police and fire memorabilia auction site; All Nations Stamp and Coin (www.downtownstamps.bc.ca) buys, sells and trades international postage, coinage and banknotes; BeerAuction.com specializes in "breweriana"; and wfpauctions.com specializes in Snoopy and Peanuts collectible items.

Weblog

It's difficult to define this type of website, mainly due to diversity of models adopted for individual weblogs (also known as "blogs"). But perhaps the *Internet Librarian 2001* describes this interesting web-based genre the best: "A web page containing brief, chronologically arranged items of information. A weblog can take the form of a diary, journal, what's new page, or links to other websites."

Weblogs are to words what [the original] Napster was to music. ("Andrew Sullivan, Wired Magazine, May 2002"). In that same issue Sullivan says, "Twenty-one months ago, I rashly decided to set up a web page myself and used Blogger.com to publish some daily musings to a readership of a few hundred. Sure, I'm lucky to be an established writer [he also writes of *The New Republic* and *The New York Times*] in the first place. And I worked hard at the blog for months for free. But the upshot is that I'm now reaching almost a quarter million readers a month and making a profit. That kind of exposure rivals the audiences of traditional news and opinion magazines."

To give you some ideas on how you can turn a profit with your blog, visit Sullivan's weblog — Andrewsullivan.com. Also click on the "Info" button on the left side of the page and check out his Media Kit.

The key to the popularity of a weblog is the person or people producing it. Since weblog readers often develop relationships with the weblog author(s), interaction between reader and author is inevitable. Good weblog examples include Eatonweb.com, Scripting.com, Gizmodo.com, and Angst-identprone.org.

But you don't necessarily need to establish a new website to attract a niche audience of readers. Since weblogs are ideally suited to interaction between people sharing special interests, some e-commerce businesses may want to consider adding a weblog to their website. A weblog's capacity for information dissemination and feedback potential can tap into the buying power of a blogging community. For example, Greg Reinacker has a weblog on the Reinacker & Associates website (www.rassoc.com), although for some reason there is no link to the blog on the home page. The only way to reach the weblog is to type in the url: www.rassoc.com/gregr/weblog/default.aspx. Another website that has successfully incorporated blogging into its website is The Caestecker (Wisconsin) Public Library (www.greenlakelibrary.org). Also check out how Redmonk.net has incorporated blogging into its content offerings.

Peer-to-Peer Site

Some readers may be toying with the idea of sitting up a peer-to-peer (P2P) site. The way the Web was originally set up, the website owner posts content on a server (referred to as "web server"), and the audience connects to that server via a web browser to view the content. To interact live with other users, everyone connects to the same server at the same time. With P2P, however, the computers of individual users are connected together directly — no central server is necessary.

Still, a website is often used as an adjunct to a file sharing network to promote the network and to provide customer service. Peer-to-peer file sharing opens a whole new range of business opportunities. The original Napster (versus the new subscription-based Napster) brought P2P to the forefront, although the original Napster wasn't a true P2P business model — it operated around a central server. However, it is noted that the old Napster's central server concept also was the cause of the business's eventual downfall. (At its peak, the old Napster was perhaps the most popular website ever created — in less than a year, it went from zero to 60 million visitors per month.)

Note: As discussed previously, the new Napster doesn't use P2P technology; rather it is a typical e-commerce site, with a hybrid subscription model that delivers downloadable music.

There are many alternatives to the Napster P2P model. One of the more interesting of these alternatives is Gnutella. Unlike Napster, Gnutella and its variants aren't software at all; rather they are communication protocols similar to the common gateway interface (CGI) used by most web servers. Any software that implements a Gnutella-like communication protocol can communicate with other Gnutella-enabled software applications.

There are other P2P technologies available; examples include FreeNet, Publius, Yaga, and others in various states of completion. All are aimed at using P2P technology to allow the free exchange of information unencumbered by censorship and unmediated by the collection of market data.

It may be difficult to come up with a viable, profit-making business model for this type of P2P network. However, visit sites such as Bearshare.com, Livewire.com and Swapper.com, for examples of how current P2P file-sharing networks use websites to produce income.

✪ CONCLUSION

Sandra Morris, vice president and chief information officer of Intel Corporation, gave a keynote speech at the Oracle Open World 2002 in Copenhagen, Denmark, where she put the dot-com bust into perspective by reading an interesting story.

The economy is stuck in the doldrums thanks largely to the broken promises of technology. Dazzled by seemingly limitless returns, bankers had funded hundreds of companies all going after the same dubious markets.

Heedless, individual investors clamored to get into the stock market, driving stock prices to unheard of levels. Sound familiar yet? Soon, the overheated market crashed, turning the new heroes of business into goats and scoundrels. Now disillusionment reigns and nobody knows what's going to happen next.

The year — 1850. The technology — the steam locomotive. The country — England.

Morris goes on to say that "Those goats and scoundrels of England, right? The parallels with the Internet are very strong. Railway mania is what this period of time is called. Hundreds of companies collapsed, tens of thousands of investors lost their money. But what happened after that breakdown was phenomenal growth for the next hundred years in the railway industry in Great Britain."

The feeding frenzy surrounding e-commerce may be new to this generation, but the same patterns of behavior occurred in the development of earlier technologies, including the steam engine, telegraphy, automobiles, airplanes, and radio. Similar to those culture-changing technologies, many lessons were learned during the e-commerce gold rush. Sure, we lost our way a bit, too many thought everything "e" was a magic bullet. But the technology that fueled that initial Internet bubble is what fuels the successful businesses of today. So, in the long term, e-commerce changed how business is conducted — and it is here to stay.

Chapter 2

Designing Your Website

Paying attention to simple little things that most men neglect makes a few men rich.

Henry Ford, Sr.

A successful e-commerce business requires a well coordinated plan that takes into account, among many other elements, design competency, programming abilities (transactional and database), server configuration, public relations, and sales and marketing abilities.

However — and even I can't quite believe I am going to say this — e-commerce is as much about selling as it is about technology. So, even though this book provides tons of advice and information on technology, just as in a brick-and-mortar store, the selling environment, i.e. your web pages, are what will make or break your e-business. You can have the best product/service in the world; but if your website doesn't provide intuitive navigation, offer an easy-to-use process for ordering and fulfillment, and maintain high standards of quality control, you will never achieve consistent customer satisfaction. Thus, your web-based business will eventually fail.

A website is an infinite number of web pages connected by a common theme and purpose. A good design is important to provide your customers with easy access to all of your website's pages. Careful consideration of the numerous design possibilities for your website is essential. Note, however, that while this book provides you with the tools you need to make the right decisions when it comes to designing your website for the e-commerce world, it does not provide HTML and programming tips.

There are many good books and websites where you can find the various HTML and programming information you might need, utilize them, if necessary.

✪ WHAT'S YOUR LINE?

Ask yourself if your product/service will translate to a virtual market, i.e., the Web. If the answer is "yes," decide which website model you want to build and consider what your website is going to offer. Then build the website and expend your efforts in driving customers to that website. If you are an entrepreneur find your niche market, design your website for that market, and then let your customers know you are open and ready for business.

If the answer is "no," a website still may be a necessity — due to the growing power of web customers who expect the vast majority of all businesses to have some sort of a web presence. Build a good brochureware site that will drive customers to your brick-and-mortar business. Use that website to provide detailed information about your company, your products and services; as the first tier in your customer service support efforts (FAQs, online manuals and other documentation, knowledge base, etc.); and perhaps even incorporate a forum for your customers to converse with each other about your products and services.

✪ TARGET "YOUR" CUSTOMERS

"Know your customer" is an oft-used but apt phrase — you must decide on your target market base before embarking on even the early design stage. Here are some reasons why. If your customers are located outside of North America, you will need to place a Comment Tag above the body of your website's home page declaring your site as a public document. If you omit the Comment Tag your site probably will not be indexed as a public document and no one outside North America will be able to find it. If your niche market has an international base, you must consider how you will provide translations, how you will handle the monetary exchange problems, and how you will deal with shipment issues. After all, it's the customers for whom you are building your website, right? This is just the tip of the iceberg — do your research so you'll know the issues you need to address *before* designing your website.

What model will your site emulate — a brochureware site, an online store, an auction site, an advertising, or subscription site? It could even be a combination of one or more of the website models discussed in Chapter 1.

What will drive your potential customers to your website? Your website's home page must clearly describe what you are offering and why your customers would want it. You must craft your online offering so that the products/services meet the wants and needs of your targeted customer base — just like a conventional business. And like a brick-and-mortar business, you must determine what price the market will bear and what your profit margin will be.

✪ YOUR WEBSITE'S BLUEPRINT

One of the most difficult parts of building a website is deciding exactly what to build. There are many e-businesses out there with ill-conceived concepts and laughable revenue models. To avoid these crippling mistakes, you need a clear vision of what you want to accomplish. To help, consider the following questions.

* What are the objectives of your new website?
* How will the new website produce income?
* What makes your new website unique?
* How will you ensure on-time delivery?
* How will you manage and maintain your website?
* How will you convey your trustworthiness and the high quality of your product/services?
* If you have a brick-and-mortar business:
    - How will you use your website to drive customers to your offline business and at the same time to provide an e-commerce alternative?
    - How will you combine your dot-com seamlessly with your traditional business into a new incarnation called "click-and-mortar"?

Your answers are the basis for the next step in building a successful website, which is the development of a comprehensive e-commerce business plan, that plan will serve as your website's blueprint. Use that blueprint to lay out the strategy needed to implement the technology that is necessary to gain the most leverage within your current or planned business model. Each aspect needs to be carefully coordinated — technical issues, content, marketing, front-end design, infrastructure, software, and, of course, sales.

There are three essential elements to a website blueprint: a storyboard, a site description, and website content. Successful websites are the ones that manage, through the proper utilization of a blueprint, to combine content, communication and marketing features within a fast-loading, easy-to-use and interesting home page that runs on a robust and scalable infrastructure.

In addition, you must establish a realistic budget and a timeline, with milestones

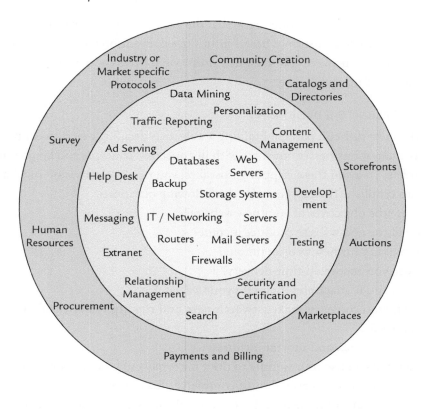

Figure 1. The E-Commerce Space. The inner circle contains your "back end" — the hardware, operating system, and most servers. The middle circle contains the applications and tools that give functionality to a website. The outer circle comprises the human elements.

clearly defined. As you define your site through your blueprint always be aware of the compatibility issues which must be considered throughout the decision making process — the extendibility and scalability of all the hardware, software, and connectivity decisions.

This can't be stressed enough — the same principles that apply to establishing a successful business in the traditional world also apply in the 24-hours a day, 7 days a week (24x7) e-commerce operation. Moreover, like conventional businesses, great websites can take months to plan and build. Defining exactly what to build, then deciding how to build and market the site is the difference between success and failure.

Storyboard

A storyboard is a tool used in the production of multimedia, video, and film projects to show a frame-by-frame picture sequence of the action. In this book, however, the term refers to a non-graphical representation of every web page — the screen elements and their operations — which, when taken as a whole, constitute your website. Just as an outline helps to organize your thoughts before you write a paper or report, storyboards help to organize a visual production such as a website. By using the storyboard process, you can design your website while clearly envisioning all the possible paths that a customer might take.

Thus, your storyboard is the visual representation of how your website will look to your customers. A good, well thought out storyboard will enable you to marry the goals and priorities for your website into a good design.

Here is how: With the proper storyboard, you can map out the progression and relationship between individual web pages. It lets you visualize how each page will work within your website before you start building it.

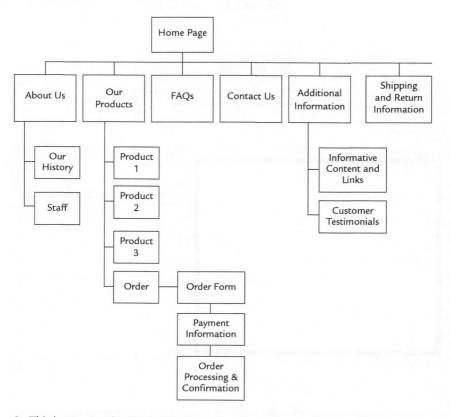

Figure 2. This is an example of a simple storyboard that might be used to design a small online store.

While tedious, creating a storyboard will save time, money, and many sleepless nights. Map out every step of your design process so that each detail can be tested, measured, and validated.

Very detailed storyboards might include an overall site diagram that shows the website on all levels: major areas of the site, secondary areas, etc.

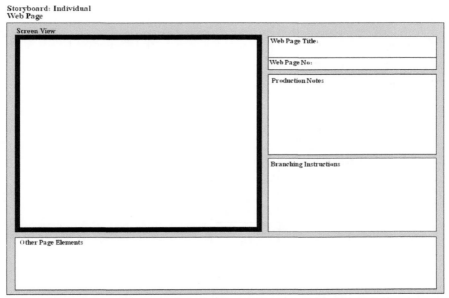

Figure 3. All Storyboards should include a basic layout of each individual web page.

To begin the storyboard process, generate a visitor-centric navigational scheme that defines the type of pages and content needed to provide your website with the necessary design elements. Take your "home" page for example: Using a single sheet of paper, describe the buttons, links, and key components that your customer should see when he or she first opens the link to your website. Then every other web page should be constructed in a similar a manner. At each step of the process, incorporate your customers' wants, needs, and perspectives. And remember that every layer of your website either precedes or supports specific choices a visitor makes, so your website's design must make sense to your visitor so you can turn that person into a repeat customer.

As you lay out the storyboard, there is one essential question you should keep in mind: What's the plot? In other words, why is the visitor here — what does your visitor want? Don't forget that many times a visitor will not reach your website through the front door, i.e. your home page, so consider all contingencies. Determine as you outline each web page: What do I want my visitors to do at this point, what do I want them to feel right now, where do I want a potential customer to go next, and how do I make it easy for them to get there?

Your final storyboard should allow your visitors to enter your website at any point — the "about us" page, the "check out" page, the "privacy policy" page — and to know where they are and to understand how they can get where they want to be quickly.

Here is a set of suggested guidelines to keep in mind as you create your storyboard.

* The storyboard should be legible. It can be created using pen and paper and does not have to be precise, but if using outside help for the design stage, it must be clearly understood by those people.
* The storyboard must be complete. Every page should be represented and every element on the page should be explained before actual design work is started or any of it programmed into HTML.
* Every design and layout element to be included on each page should be noted in the storyboard —
   - Headings
   - Text objects/blocks
   - Links/Buttons
   - Graphics images (photos and other arts).
* The typeface and print size of the headings should be exactly as they will appear on the final web pages.
* The number and the function of the buttons should be clearly indicated on each page.

* Links between pages should be clearly indicated using arrows.
* Each graphic image should be noted with a box identifying it as a graphic image with a short note describing the content.
* Web pages should be numbered for easy reference.

Think in terms of who will see your web pages. Perhaps it will be a potential customer who has no idea who you are or what you have to offer. Thus, the best way to layout your storyboard is to track the path of a hypothetical customer, with branches at every decision point — including those made by the customers and those made by the system. Have a meeting with all of your staff — sales and marketing, customer service, public relations — not just the website staff. Get everyone's input; cover all the possibilities. For example, in the purchasing process:

* Does the site require customer registration before the purchasing process can begin?
* Is there an option to skip registration but to allow the purchasing process to continue?
* If a customer wants to change or to remove an item from the shopping cart, is it easily done?
* At what point does the credit card authorization take place?
* Is there a confirmation page that also provides an order tracking number?

Decisions made at this point must *not* be rushed. Time is needed to study, absorb, and totally understand what's required to implement the most creative ideas — the ideas that will make your website stand out from the crowd.

Use your storyboard as your guide throughout — design, build out, and beyond. Storyboarding helps not only to improve site navigability, but also to develop content and web copy. Furthermore, if you hire a web designer to design your web pages, or a web architect to oversee the entire build out phase, the storyboard will provide them with the details necessary to provide you with exactly what you want.

The author realizes that both layout and design are subjective topics, but to make the best first impression design a stylish page with your content laid out in a logical manner. Use a consistent theme in the colors, styles and fonts throughout your site.

Site Description

If you are designing an extensive website, you need to provide a detailed explanation of workflow, data flow, and other items that may not be readily apparent in your storyboard. That's where a site description comes into play. A site description explains how the site functions from web page to web page or section to section. This is a must for complex sites, such as websites that include an auction element or websites like

3Com.com or Healthtex.com. That is because oft-times such websites have people who are not intimately familiar with the website's design elements and infrastructure doing the programming to support the website's more complex elements.

Website Content

Now is the time to begin thinking about exactly what content you want on your website. What digital art (e.g. photos of products) will be needed? Is there to be written content? If so, who is the author and how will the content be delivered — MS Word, Adobe Acrobat, ASCII text, etc.?

You should have the initial content ready for the designer(s) (which could be you) while the website is still in production mode. This is so you can be sure that your content will work perfectly with the overall design elements when it comes time to launch your website (and when subsequent design and content changes are made thereafter).

✪ DESIGN

The successful website starts with a home page that is attractive, easy to understand, and fast loading. Think of your home page as the cover of a good book — it should entice the customer to look deeper into the site (book) and return to it often as a resource. Another way to put it is that your home page, the first page the online consumer will see, is like the window of a store. It is your showcase, storefront, and calling card — all rolled into one. Online, your competition is just a click away — careful design and targeted content are important guardians against customer defection.

Your website's design and content will greatly influence your customers' perception of your business, which will, in turn, affect their purchasing decisions. Your pages should be laid out in such a manner that navigation through your site is intuitive and stress-free — so much so that your customers develop a comfort level in doing business on your site. How do you manage that? Read on!

When designing your site, there are certain categories of rules or guidelines that you should follow. The acronym SPEC can be used to help you to remember the key categories:

Stickiness and traffic generation
    Content
    Search engines
Performance
    Speedy downloads
    Tables

Ease of use
    Site Navigation
Content visibility
    Viewable Site
    Frames
    Java
    Plug-ins

Stickiness and Traffic Generation

A sticky website entices a visit to stay within its pages a bit longer than they otherwise might; over time that leads to a familiarity with your website. The more familiar some-one is with a business — online or off — the more comfortable they are when it comes to making a purchase. Thus, a sticky website is one that keeps visitors not only within its web pages, but also keeps them coming back for more. This dynamic is created with a mix of good content and good design.

Content

An important sales adage is — CONTENT IS KING. "Content" is your website's offer-ing — the product, the graphics, the marketing material, banner ads, i.e., everything that makes up the pages of a website. Good content gives a website a high "stickiness" rating. In other words, good content entices customers to stay within your website, and encourages them to return to your website time and time again.

Your website has taken the first step towards being a success when you follow the Internet's golden rule — Provide Useful Content. Independent of which e-commerce business model you adopt, the content must be presented in such a manner as to draw a visitor's immediate interest and even more importantly, it must turn that visitor into a loyal customer. Your content should include all the information necessary for a customer to make intelligent purchases in an easily accessible way.

By keeping the content of your website fresh and new, your customers will be more likely to "bookmark" your website, or at least a specific page within your website. Curiosity is a powerful lure and customers will come back to your site repeatedly just to see what is new. It's the useful and up-to-date information that will keep your cus-tomers coming back time and again.

Search Engines

Make a list of the top ten terms that your customers could use to search for your web-site when using a search engine. Then incorporate these words in your web page content, i.e. make sure your web pages include text relevant to those ten terms. The majority of search engines do not index by keyword submissions alone, they send out spiders to crawl your site to check that the keywords you submitted are relevant to the content contained within your website. Why? Because disreputable website owners, especially pornographic and gambling sites, submit numerous keywords that people use every day in their search criteria that have nothing to do with the content of the website. These same unscrupulous owners will also insert unrelated words and phrases into their meta tags. (See Chapter 15 for a full discussion on how to design your website to obtain optimal search engine ranking.)

Performance

Long download times are unnecessary and unprofitable. Making a potential customer wait for your website to download is a surefire way to increase your competitors' bottom line, not yours.

There are many reasons why a page may load slowly, e.g. the size of the pipe to the Internet, the traffic hitting the web-hosting service and/or the server hosting your website, the robustness of the web server, etc. But at this juncture, you just need to ensure that your website's design is not a contributing factor when a customer experiences a slow download. So keep your home page less than 200 KB in size. By doing this, your website will load in less than 20 seconds with a 56K modem.

If you must display graphics on your home page, keep in mind the different graphic formats that are available, each with its own qualities and capabilities, and what is best to use in specific situations. Web-based images consist of two basic types: those captured from nature and stored in digital format, and those created entirely on the computer.

Most web-based images use "indexed color" which is only 8 bits (one byte) of color per pixel. This means that the image can display only 256 colors. Don't panic — it isn't quite as bad as it sounds — you can choose your 256 colors from a huge palette of 24-bit colors. If you pick the right colors, even a color photo can be made to look presentable on your website. Many programs such as PhotoShop and Paint Shop Pro will let you reduce the number of colors (color depth) in an image, while selecting the colors closest to the original.

GIF and JPEG (also known as "JPG") are the most commonly supported formats throughout the world. "GIF" stands for "Graphics Interchange (or Image) Format." CompuServe developed the GIF format so that its subscribers could send image files to each other and the images could be viewed on different kinds of computers. A GIF is good for images that have solid colors, text, and line art. A GIF can be used to represent images generated by drawing programs used by computer artists. However, a GIF does not compress photos very well; especially images that have subtle texture or color gradations, or that are 16- or 24-bit color.

JPEG stands for "Joint Photographic Experts Group." (In the DOS world, JPEGs were called "JPGs" because DOS filenames were restricted to having only a three-letter extender.) JPEG is perhaps the best format to use for photographs since it supports full 24-bit color.

To sum up: Web-based GIFs are better for solid and flat colors, exact detail, sharp edges, black and white images, images with transparent areas, simple animations and small text. Use JPEG for images with continuous tones, such as photographs or images with gradient fills.

But also remember that the higher the quality/resolution of the graphics, the larger their file size. The larger the file size, the longer it will take for the web page to load.

Note: One solution to the size versus resolution issue is to use thumbnail JPEG images on the home page and then link these to corresponding full-size images on another page. This way you give the customers what they want — detail and a fast load time. A good example of an e-commerce site using this technique is Artcut.com.

Tables

Using HTML tables on a web page allows the organized and specific arrangement of data. The data can be text, images, links, forms, form fields, other tables, etc., arranged into rows and columns of cells (individual units). Tables let you control the look of your website by breaking your pages into precise segments while controlling the placement of text and graphics. You can create columns and grids that can contain images and text. Cells can be utilized as templates or style sheets to give a uniform look and, through use of color, add visual contrast to your website. Be careful though, if you use colored cells in your table, some browsers might not display the cell in color unless there is text or an image in it. For example, older browsers can only see the background color described in the <BODY></BODY> tag. Also Netscape's handling of empty table cells give web designers fits. That is because Netscape browsers have a well-known bug that prevents the browser from displaying empty table cells. There is a "work around," but it means the designer is required to add a bunch of code to the table if you want your empty cells to display background colors when your website is accessed via a Netscape browser.

There are also other problems with tables. They may load more slowly than plain text since some browsers must "place" the items in a table, which means the table won't be shown until all the text and graphic items have downloaded. However, you can

Original layout of website using tables

As viewed by visitor
using a Netscape browser

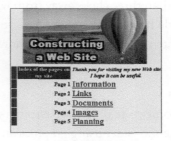

As viewed by a visitor
using an Internet Explorer browser

Figure 4. Use tables to align objects and accentuate with color and image backgrounds. However, when using tables be aware that what your WYSWYG web design software shows what you and what visitors actually see may be different.

mitigate this problem by breaking a long table into smaller tables and specify the height and width for all the images. This fix allows the browser to size the table before the images, which in turn, results in a faster download. Another advantage to using a series of smaller tables instead of one long table is that it is easier to change the page's design in the future.

You also must deal with the fixed-width problem. When you add a variable width table, the horizontal dimensions readjust with the browser width. A fixed width guarantees the final appearance. But the variable width can take better advantage of the situation if the browser has a larger width setting to begin with. Use trial and error testing to find the optimal combination of fixed and variable widths for the different parts of the table.

Some WYSIWYG (what you see is what you get) web page editors (this is software) have problems with tables. Be careful, if you are doing your own design work — what your editor displays on the screen may vary greatly from what you see in a browser (See Fig. 4).

Test your table-based pages with a variety of browsers and don't forget to use different browser widths. Also test the page scrolling function by using your mouse to pull down the arrow on the bar (which your browser automatically brings up on its far right side of your page) to see if the page jumps rather than scrolling smoothly.

Ease of Use

A simple, easy-to-understand navigational design ensures your customers quick and effortless travel through the multiple pages on your website. Your customer should never be more than three clicks away from what they want to find. Without fast, intuitive, and simple navigation capabilities, your customers will not take the time and effort to navigate your site, regardless of how good your content, product and/or service might be.

Site Navigation

Design your home page to allow customers access to all areas of your website from your home page. Consider using graphics and image maps — a clickable picture (when you place your cursor on it, the cursor turns into the "link select cursor") — as an attractive means of navigation. But remember the users who surf the Web using text-only browsers by also inserting text links (a typewritten description not dependent on an image) at the top or bottom of each web page.

As you drill down into the site, continue a uniform navigation scheme, i.e., the customers can go to the same position on any page to perform a specific function. Don't forget to institute "targeted" text links (i.e. text that you can click on and be transported to a specific section of the website), especially in pages that are long or divided into topics or resources. By doing so, you allow the customer more easily to find whatever they may be looking for. Targeted links can be an expedient form of navigation, supplementing the scroll bar.

Structure your site's design to support future growth. But even when designing for future complexity (e.g. the addition of an auction section or adding a shopping cart), always keep the customer's view in the forefront. For example, don't make the mistake of asking your customer to remember a certain product ID or code when it comes to filling out the order form — keep it simple.

Avoid orphan pages — pages that although there is a link leading to the page, the page offers no link to leave the page. Such pages give a potential customer a choice that might lead out of your website, because to continue their search they must either hit the back button on their browser, or close the page and go elsewhere. To avoid this dilemma, always consider all possible navigational paths a visitor might take and then ensure that there is a series of relevant links available on each page providing your potential customer with some very good reasons for him or her to stay in your website.

Content Visibility

Design your website so that it is technically accessible to the greatest number of people. Just as customers come in all shapes and sizes so do the equipment and the software that they use to access the Web.

Viewable Site

Test your design with as many browsers (including their various versions) as you can find — Netscape, Microsoft Internet Explorer, Macintosh, Opera, AOL, and a text-only browser such as Lynx. Don't forget the customers that surf with their browser's "turn off graphics" option activated. In other words, make sure the technologies you select can accommodate the many browsing options your customers will be using.

Frames

Also called framesets, frames are a programming device that divides web pages into multiple, scrollable regions. This allows you to present information in a more flexible and useful fashion. Of course, frames have their own set of problems. A browser's back button can produce unexpected results, particularly if the user is working with an old browser, such as Netscape prior to version 3.0.

Visitors who have problems with their sight or are otherwise physically impaired may be using text-to-speech software that reads aloud web pages. Frames confuse such software.

Even for the non-physically challenged, a cursor may not work with a framed site unless you actually click in the frame you want to scroll.

In addition, frames can make it more problematic to print. For example the Princeton Online website, which is designed around frames, actually has a web page devoted to helping its visitors print information available on that website (see Fig. 5). But, even worse, framed websites may be invisible to certain search engines and directors (such as Yahoo!). Frames increase the file size and the number of total words that make up the website, thereby decreasing keyword weight and perhaps causing an adverse effect on your website's search engine listing. Also, when customers are brought to your website via a search engine, they sometimes won't enter through the front door, i.e., home page, and therefore can't see the frame that would normally be holding the page.

Exercise caution if you choose to offer links to other web pages within a framed page. The linked page can accidentally load within the framed page on your website;

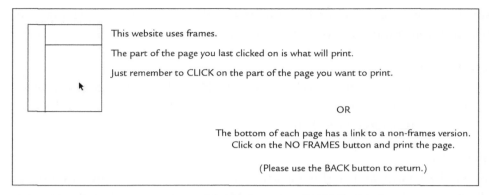

This website uses frames.

The part of the page you last clicked on is what will print.

Just remember to CLICK on the part of the page you want to print.

OR

The bottom of each page has a link to a non-frames version.
Click on the NO FRAMES button and print the page.

(Please use the BACK button to return.)

Figure 5. The Princeton Online website found that so many visitors experienced problems when trying to print information available on its frame-based website that it crafted a "print help" web page.

particularly if the correct code is not inserted (such as _blank, top, parent, self, or your own designated frame) to keep external sites from loading within your frameset. This can be confusing to the visitor as well as raising the possibility of copyright infringement, since the user may think that the information appearing in your frame is your information.

When advertising one particular aspect of your site in other media, simply providing a main URL address is no longer good enough. In the case of a framed site you must give the public additional instructions about how to find the frame and the page that they want. If you give out that page's address alone, the rest of the frameset becomes inaccessible.

Conversely, if another site wants to link to a particular page on your site, they're out of luck. Frames generally restrict external-to-internal links only to the home page, unless you once again want to find yourself on a page minus the rest of the frameset. In other words, if a website wished to link to an internal page of your website (like your FAQ section) it couldn't — it would only be able to link to your home page.

Creating a non-frame site is the best approach to make it accessible to the largest number of users. All things considered, it is better that you design your website using tables, which offer some of the same functionality as frames but with fewer limitations.

Java

Java is a high-level object-oriented programming language (a set of grammatical rules for instructing a computer to perform specific tasks) developed by Sun that is all of the following:

* Simple
* Architecture-neutral
* Object-oriented
* Portable
* Distributed
* High-performance
* Interpreted
* Multithreaded
* Robust
* Dynamic
* Secure

All of the preceding buzzwords are explained in "The Java Language Environment", a white paper written by James Gosling and Henry McGilton. You can download a PDF version of the paper from http://java.sun.com/docs/white/.

Note: Programming languages, while simple compared to human languages, are more complex than the "machine language" computers actually understand. A computer operates using binary numbers, i.e. it only understands ones and zeros. However, humans have a hard time with the long strings of ones and zeros that are the heart of machine language programs. We work better with decimal numbers and words. That is where programming languages come into play — they are the bridge between the word-using human and the binary-using computer.

Probably the most well known Java programs are Java applets. These programs, written in the Java programming language, can be included in an HTML page, much in the same way an image is included. When a customer uses a Java technology-enabled browser to view a page that contains an applet, the applet's code is transferred to the customer's system and executed by the browser's Java Virtual Machine (JVM).

With java applets you can add to your website anything from a small animation to a sophisticated program that displays a 30 second movie. The applets can either run in the same HTML page or in a popular (but sometimes annoying) "pop-up" window that opens as a separate window when surfing a website. To see examples of sophisticated applets that runs within the same HTML page visit www.really-fine.com/reallyfineart.html and www.crownnet.co.uk/portfolio/vr-appletpano.htm.

For other examples of what can be done with Java Applets, go to Jars.com. Sun also offers a number of free "tried and true" Java applets that you can use on your website. To see what's available go to http://java.sun.com/openstudio/index.html; also visit http://java.sun.com/applets, which offers a number of other Java resources.

Although Java applets, if built correctly, can enhance the functionality of your website, they could become a crippling factor if used extensively. This is because Java applets may take extra time to download and some applets can be very demanding, which may cause problems for the customer's computer (especially if the machine is an older or low-end model).

Furthermore, if you are trying to reach as many people as possible, keep in mind that not everyone viewing your site will have a Java-enabled browser or have the "Java-enabled" feature turned on. So, if you would like to use applets, use them only when it is not important that everyone coming to your site have the ability to view the information contained in the applet. Another suggestion — if possible, never use them in your home page. Also, don't design your entire website so that it can only be viewed with Java-enabled browsers. Why? Because, when customers who have browsers that don't support Java come to your site, they see — instead of your website — an irritating message stating that the site requires a Java-enabled browser.

Also, when deciding on whether or not to use a pop-up window (i.e. a java applet), keep in mind that some customers find pop-up windows disconcerting.

Plug-Ins

Generally, plug-ins are software modules that run on the viewers' local machine and add to the functionality of an application. Typically, web browsers use plug-ins so that they can display a wider range of formats. For example, an Acrobat plug-in is used to view documents in Acrobat format (PDF documents). Most video and audio formats require a plug-in to be viewed or heard. For instance, a QuickTime plug-in is needed to view movies in QuickTime format. When a customer tries to view something that requires a plug-in, which their browser does not support or they have not previously installed, they will get a message asking them whether they would like to install the plug-in.

Plug-ins are free for the most part, but downloading and installing them requires some sacrifice on the part of the customer. Some plug-ins are quite large taking a considerable amount of time to download on a 56K modem, and space availability may become an issue for the customer — a minimum of 3MB is usually required for a plug-in. One more consideration is that many plug-ins are not backwards compatible with previous versions. This means that even if a user has previously installed the required plug-in, there is no assurance that it is the correct version. For example, if a customer has the Macromedia Flash 3.0 plug-in and you use Flash 4.0 on your website, the customer will need the Flash 4.0 plug-in to view your site, since the 3.0 version will not work.

Thus plug-ins can create a great barrier between your customer-base and your content. In fact, a number of recent studies show that less then 10% of the Web population use plug-ins. Furthermore, these same studies also indicate that many potential customers might be intimidated by plug-in based content. Yet: If you know that your target market is technically astute, plug-ins may help to put your website ahead of its competitors.

If you are unsure about the technical prowess of your target market, and you want to, say, offer an important document in PDF format, then also provide a text only version that is easily viewable by all customers.

If you build your site with Macromedia technology, understand that it will be viewable only if the customer's browser supports the *exact* version your website is using. Otherwise, the customer is required to first upgrade his or her Macromedia plug-in before they can view that content. It is strongly recommended that you also provide an HTML-only version so that all of your customers can view at least one version or the other of your website. The author isn't against Macromedia technology — it allows you to design a great looking website. But at the same time, the reader must understand that although Macromedia allows you to build a eye-catching site, it doesn't do much good if your target audience cannot access it.

For an example of a website that incorporates all of the advice set out in this section, visit the website of Brown Beattie O'Donovan LLP, (www.bbo.on.ca).

✪ CONCLUSION

You can design a website that is brilliantly complex, employing all the latest technology or you can choose to design a simple site without sacrificing attractiveness or efficiency. If you keep your customers in mind and design your site to accommodate the lowest denominator (technologically speaking) in your customer base, you will have a fighting chance of succeeding on the World Wide Web.

Don't skimp on any of the essential elements to your website's blueprint. All three — storyboard, site description, and website content — are indispensable. For every hour you spend planning and getting all the details right, you save yourself the cost and time of at least three days' of remedial tinkering and development. Every decision you make now will define and limit the future growth and evolution of your website.

You are now ready to build your website.

Chapter 3

The Devil is in the Details

It isn't so much that we have a new economy, as we have a new understanding of the importance of technology in the economy.

—Paul Romer, Professor of Economics
Stanford University

You've chosen your e-commerce model and found the perfect products/services to offer on your e-commerce site. You've also thoughtfully planned your website. Using your blueprint and storyboards you've completed the design of your website. It is now time to extend everything to the Web.

The basic e-commerce website should:

* Store any number of products that have been selected by the customer prior to the actual processing of the purchase. This system is normally referred to as a "shopping cart," processing is usually referred to as "check out."
* Provide a secure server with SSL encryption for transactions, email transmission, and storage.
* Accept credit cards and offer automatic, real-time processing. But offline processing via an encrypted email form is also a viable option if you choose to forego the following options.
  - Allow the customer to leave the site, return at a later time, and still find past items in their shopping cart.
  - Allow cross selling, i.e.; offers a similar product to the one that the customer is interested in, if the chosen product is unavailable.
  - Provide processing status though a numbered tracking system.

Add to this list: acquiring a domain name, a merchant account, and a digital certificate, and you are in the e-commerce business.

✪ DOMAIN NAME

Let's first look at choosing a web address a/k/a domain name. A domain name is your web business's cyber address — it's also known as your site's URL (Uniform Resource Locator). You've no doubt seen the many "dot-com" advertisements — www.[name].com (or .org or .net). That is a domain name or web address. It's how the public will find your web-business unless you have opted for the cyber mall concept.

While you can choose just about any combination of words or numbers for your domain name, we recommend a catchy, easy-to-remember name that can serve to quickly evoke your business and/or the products and services it offers. Come up with several options.

In your quest for the perfect domain name remember:

Your online business depends on the customer correctly typing your URL — the shorter the better. And please, don't put your entire name or your company's name in the address. No one wants to input www.the-one-and-only-genuine-original-widget-company.com. Find something simple.

If your brick-and-mortar business has a well-known name that is already branded, re-enforce that brand online, don't create an entirely new "web name." Remember that brands are expensive to promote, particularly new ones.

Think twice before you use "web" or ".com" in your name. Yes, we know .com is probably part of your URL but it is not necessarily part of your name, which will, by necessity, be branded. Why? Because technology and the growth of the Internet are moving at breakneck speed and "web" and ".com" will, in the future, appear stale and dated. In the new world of fast moving technology your business should always present the image of being on the cutting edge.

Competition for rights to domain names has exploded. Many people and companies have registered not only the domain names they use, but names they think may be valuable in the future. Check the Network Solutions' WhoIs directory (www.networksolutions.com/cgi-bin/whois/whois) to see if your chosen domain name is available. If your ideal domain name isn't available, you might consider contacting the owner of that particular domain name to try to purchase the rights.

Once you have chosen your domain name / web address / URL, the next step is to register it so you can have an exclusive home for your online business. Registering your chosen domain name with a domain registration site ensures that you "own" that specific web address; at least as long as you continue to renew your ownership by paying the required annual fee. The process itself is easy, but there is much you need to know to begin this process.

First, understand the Internet's system, known as the "Domain Name System," which keeps track of the millions of computers that are connected to its byways. The Domain Name System (DNS) allows data packets to find their way to their destination.

Every computer on the Internet has a unique address called an "IP address" (IP stands for "Internet Protocol"). But that address is a rather complicated string of numbers, which is hard for the average person to remember.

THE IP ADDRESS

A global, static IP address is a unique 32-bit binary number representing just one resource on the Internet, at any given time. These global addresses are assigned by one of three Regional Internet Registries (RIRs) worldwide that collectively provide IP registration services to all regions around the globe. The American Registry for Internet Numbers (ARIN) covers IP addresses in the geographic areas of North America, South America, the Caribbean, and sub-Saharan Africa. APNIC (Asia Pacific Network Information Center) and RIPE (Reseaux IP Europeens) cover their own specific regions.

An IP address's 32-bit binary number is actually made up of four bytes of information, and each byte can be represented by an eight-bit binary number or "octet." Since we find it easier to read and write numbers using digits from 0 to 9 instead of giant binary numbers consisting of just ones and zeroes, IP addresses are expressed as four decimal numbers, each separated by a dot. This format is called "dotted-decimal notation." The IP address is also called a dot address, because periods are used to separate four sets of decimal numbers, each between 0 and 255, representing all 256 possible combinations of eight bits (00000000 through 11111111). Each of the decimal numbers is called a "quad" because four of them make up the IP address. Thus, an IP address is also sometimes called a "dotted quad." Here's an example:

Dotted-Decimal Notation

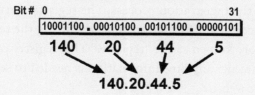

Figure 6. The theoretical decimal range of IP addresses is from 0.0.0.0 to 255.255.255.255. When translated into binary numbers, that's 00000000.00000000.00000000.00000000 to 11111111.11111111.11111111.11111111. Actually, we're restricted from using all zeroes for the individual computer (or host) part of an address (it's an address used for routing, representing the originating network) or all ones (used for broadcasting to all other computers on a network).

When you type in or click on a link such as www.microsoft.com, that domain name (www.microsoft.com) has no meaning for your computer, but its associated IP address is a different matter — it is what's used to connect your browser with the Microsoft site. Thus, when you input "www.microsoft.com," your computer sends a message to a DNS server on the Internet for the Microsoft website's IP address, and that 32-bit binary number is used to connect your browser with the Microsoft site. There are millions of computers, millions of websites, and millions of IP addresses.

It's easy to see how the DNS simplifies using the Internet through the exchange of a familiar string of letters (the "domain name") for an arcane IP address. So instead of typing 216.239.51.99, you can type, www.google.com, to reach the popular search engine website.

In the final analysis, the Domain Name System was established to provide a "mnemonic" device that makes it easier for people to remember Internet addresses.

Registering Your Website's Domain Name

When you register a domain name, you are inserting an entry into a directory of all the domain names and their corresponding computers on the Internet.

To register your new domain name you must use the services of an accredited registrar. Domain names ending with .biz, .com, .info, .name, .net or .org can be registered through many different registrars that compete with one another.

Note: InterNIC maintains an up-to-date list of accredited registrars, which can be found at www.internic.net/regist. For information on the registrar accreditation process or to lodge a complaint about an accredited registrar, visit www.icann.org/registrars. The Internet Corporation for Assigned Names and Numbers (ICANN) is a non-profit corporation charged with the responsibility of managing IP address space allocation, protocol parameter assignment, domain name system, and root server system.

The registrar you choose will ask that you provide various contact and technical information as part of the registration process. The registrar is then charged with keeping records of the contact information and submitting the technical information to a central directory known as the "registry." This registry can be accessed by any computer on the Internet any time information is needed to send an email message or to find a specific website.

The registrar also will require that you to enter into a registration contract. That document sets forth the terms under which your registration is accepted and maintained by the registrar.

ccTLDs. Some readers may also want to register a domain name using a two-letter

THE TOP LEVEL DOMAIN SYSTEM

To understand the Top Level Domain (TLD) system, you must grasp the role that the Internet Corporation for Assigned Names and Numbers (ICANN) plays as the technical coordination body for the Internet. As such, ICANN provides a global forum for developing policies for the coordination of some of the Internet's core technical elements, including the Domain Name System (DNS). This non-profit organization operates on the basis of consensus, with affected stakeholders coming together to formulate coordination policies for the Internet's core technical elements in the public interest. The policies are then implemented by the agreement of the operators of the core elements, including generic Top Level Domains (gTLDs) registry operators and sponsors, country-code Top Level Domains (ccTLDs) managers, regional Internet (IP address) registries, and root-name server operators.

Originally, the agreement to implement coordinated policies for the Internet was informal. But, as the Internet grew in commercial importance, operators and users of the Internet concluded that a more formal set of written agreements must be established. One of ICANN's activities is to work with the other organizations involved in the Internet's technical coordination to document their participatory role within the ICANN process and their commitments to implement the policies that result. These have included agreements with Network Solutions (now VeriSign), which operates the .com and .net gTLDs; the companies responsible for operating the "unsponsored" TLDs (.biz, .info, and .name); the organizations sponsoring the "sponsored" TLDs (.aero, .coop, and .museum); Public Interest Registry, which operates the .org gTLDs; more than 150 ICANN-accredited registrars; the regional Internet registries; and the Internet Engineering Task Force.

Let's look more closely at the Top Level Domain system. There are several types of Top Level Domains (TLDs) within the Domain Name System. They include:

- ccTLDs. As previously discussed, this refers to two letter country-code top level domain names. They have been established for more than 240 countries and external territories. Designated "managers" manage the ccTLDs according to local policies that are adapted to best meet the economic, cultural, linguistic, and legal circumstances of the country or territory involved.

- gTLDs. Most TLDs with three or more characters are referred to as "generic" "gTLDs." Originally, only seven gTLDs (.com, .edu, .gov, .int, .mil, .net, and .org) were created, although domain names may be registered in only three of these (.com, .net, and .org) without restriction; the other four have limited purposes. As the Internet grew, various discussions occurred concerning additional gTLDs, leading to the selection in November 2000 of seven new TLDs. Four of the new TLDs (.biz, .info, .name, and .pro) are unsponsored (uTLDs). The other three new TLDs (.aero, .coop, and .museum) are sponsored (sTLDs). Generally speaking, an unsponsored TLD operates under policies established by the global Internet community directly through the ICANN process, while a sponsored TLD is a specialized TLD that has a sponsor representing the narrower community that is most affected by that TLD. The sponsor thus carries out delegated policy-formulation responsibilities over many matters concerning the TLD.

- .arpa. This special TLD is used for technical infrastructure purposes. ICANN administers the .arpa TLD in cooperation with the Internet technical community under the guidance of the Internet Architecture Board.

For more information on the details of TLDs, go to http://www.icann.org/tlds.

country-code, known as country-code top-level domains or ccTLDs. The use of ccTLDs was introduced by Dr. Jon Postel, the Internet architect originally entrusted with responsibility for deployment of the Internet's domain name system. His objective for the DNS system was to enable local Internet communities worldwide to develop their own locally-responsive and -accountable DNS services, and to encourage all parts of the world to "get online." That original initiative has grown into the ccTLDs used today to document various countries' (and territories') relationships with ICANN. Examples of such ccTLDs include .ae (United Arab Emirates), .au (Australia), .ca (Canada), .fr (France), .jp (Japan), and .uk (United Kingdom). Such registrations are administered by what's known as "country-code managers." To identify the manager for your specific country-code, and for information about ccTLD registration requirements, see the IANA ccTLD database, which can be found at www.iana.org/cctld/cctld-whois.

✪ DIGITAL CERTIFICATES

Digital certificates (also referred to as "authentication certificates," "SSL server certificates," and "digital IDs") are the key to providing customer transaction security. A digital certificate is a message sent by one party to another at the beginning of a secure Internet session. The certificate verifies the sender's identity and vouches for that person's/organization's integrity. Just as a driving license is used to validate a motor vehicle driver, a digital certificate establishs the identity of someone in cyberspace. These digital IDs hold a mapping between a user and an encryption key. This key is private to the user and only he or she can use it. Digital certificates also contain the information necessary to allow users to exchange data securely and to transact business over the Internet.

You obtain your digital certificate from an organization called a "Certificate Authority" (CA). The certificate is virtually impossible to forge because the final requirement of secure communications is non-repudiation: a message's source must be able to be proven beyond a reasonable doubt upon demand.

Technically, a Digital Certificate is a small piece of unique data used by encryption and authentication software. This digital ID establishes a user's credentials when doing business or other transactions on the Web. It does this by attaching a small file to the data transaction. That file contains: the certificate owner's name, a serial number, expiration date, a copy of the certificate owner's public key (used for encrypting messages and digital signatures), and the digital signature of the certificate-issuing authority, allowing the recipient to verify that the certificate is genuine.

Once an Internet client (i.e. your customer's web browser) requests a secure session, your web server sends the client/browser its digital certificate information, which contains the following:

* The public key.
* The certificate's serial number.
* The certificate's validity period.
* The web server's official domain name.
* The domain name of the CA that issued the certificate.

This information is encrypted using the private key of the CA that issued the certificate. Upon receiving your digital certificate information, the customer's web browser validates it by checking the following criteria:

* Validity — is it valid for the current date?
* Ownership — does it belong to the server that sent it?
* Certificate Authority — is the CA that issued the digital certificate known and trusted? (To do this, the client/browser checks the CA's own certificate, which is signed by the CA itself.)
* Public key — can the CA's digital signature be decrypted using the CA's public key. (Most web browsers contain a list of the public keys of the best-known CAs, so they do not need to search the Internet for them.)

If the certificate fails any of these tests, the client/browser issues a warning to the customer. The customer may then choose either to continue with the session or to discontinue the session.

Digital certificates can be issued in chains. For example, a large CA might issue a certificate to a smaller CA, which issues a certificate to a still smaller CA, which issues end-entity certificates to e-commerce websites. This type of procedure helps to distribute the task of administering digital certificates. When a customer's browser/client receives a certificate from a chain, it checks the certificate of every CA in the chain as described above, until it reaches the self-signed certificate of a top-level CA.

Perhaps this scenario will help the reader better to understand how a digital certificate is used in e-commerce. (Also see Fig. 8.) Let's assume a customer fills in his or her credit card details on a form at a website and then clicks the buy / send / submit order button, which causes that customer's credit card information to be sent as little bits of electronic data from the customer's PC to another computer, via the Internet (which is made up of a network of computers, wires, cables and other connections). If not careful, the customer's data can be intercepted by malicious third parties (this is why transaction security is vital in e-commerce). Such security is provided by spe-

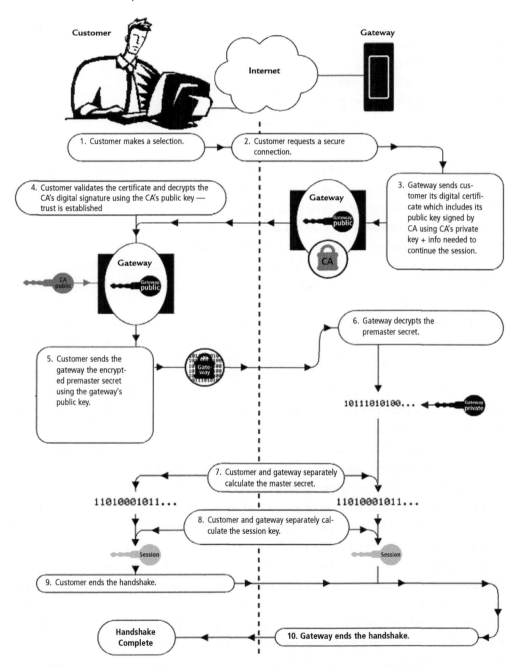

Figure 7. A typical e-commerce related digital certificate scenario. Digital certificates are one of the primary keys to effective e-commerce security because they provide the means for verifying identity.

cially-configured servers and encryption technology, which ensure that the data transferred between the customer's PC, the website, and any other destination (such as the computer responsible for processing the credit card) has been turned into a special encrypted code that is difficult to unencrypt, and thus unusable if intercepted.

SSL

Central to the digital certificate process is the secure socket layer (SSL) standard, which standardizes the way web browsers and web servers communicate with each other using encrypted data. All but legacy web browsers are set up to conduct secure exchanges of electronic information using the SSL standard. However, a secure exchange can't take place unless there is a digital certificate installed on the web server in question. The

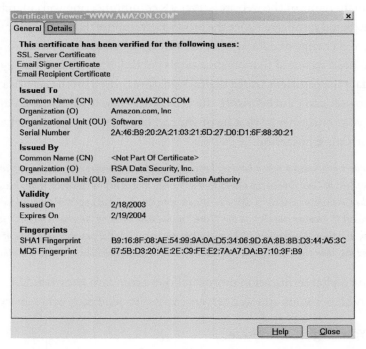

Figure 8. To see what your customer might see when accessing your website via a Netscape browser, we will use Amazon.com's secure server feature. When a customer signs onto Amazon.com using the site's "sign in using secure server option," Amazon's digital certificate is invoked. To view the digital certificate information, the customer just right clicks any place on the page. That action will bring up a menu and at the bottom of that menu is the selection "View Page Info," click it. That will bring up a "Page Info" window with several tabs at the top, choose the tab entitled "Security." Next will be a window that begins "Web Site identify verified" and in the middle of that window is a "view" button, click on that button to bring up the "Certificate Viewer." There the customer can see the details of the Amazon.com digital certificate as shown in this graphic.

visitor's web browser uses the digital certificate to authenticate the server and to initiate the encryption process. The browser uses the certificate not only as a stamp of approval, but also to access the information needed to activate the SSL process. If there is no certificate, then the browser (and thus your potential customer) can't be sure that the server is a reliable partner with which it can exchange data.

Here's an illustration. A potential customer finds your website, and selects products that he or she wants to purchase. When that customer initiates the credit card transaction, one of the first items of business is for his or her web browser to authenticate your web server by examining its digital certificate. The examination ensures that the certificate has not passed its "best before date" (i.e. it is still valid) and the organization that issued the certificate is on its list of approved certifiers. The browser may also check that the domain name of the website/server it's communicating with matches the one listed on the digital certificate. Once the browser determines that your server is approved for the secure exchange of data, it will continue with the secure transaction. Since the browser also has a record of that organization's public key, it will use this key to decrypt and to validate the certifier's coded digital signature carried on the certificate.

If the browser can't authenticate the certificate then it will issue an appropriate warning to your customer so that he or she can make the decision whether or not to continue with the transaction.

Note: Some readers may wonder how a browser knows if the CA is trustworthy. Web browsers come with their own pre-approved list of CAs. One of the checks that a browser makes before committing to a secure connection is to compare the CA listed on a server's digital certificate with its own list of approved CAs. To see what I mean, go to your Microsoft IE browser and click on the "Tools" menu, then check on "Internet Options," followed by the "Contents" tab, and then "Certificates." This will allow you to see a list of pre-approved CAs. If the CA is not on this approved list, then the web browser will invite its user to decide whether or not to trust the certificate.

This entire digital certification process follows a standard known as SSL — a process that virtually all browsers and web servers can use to send each other encrypted data.

Obtaining Your Digital Certificate

As you should understand, the secure exchange of data through SSL takes place only when the web server has a digital certificate installed. If this license to conduct secure transactions is not present, then SSL is not activated, meaning it isn't possible to provide a secure transmission of information for your customers.

There are two ways to obtain a digital certificate: Your hosting service owns a digital certificate that you might be able to use for a fee or you can purchase your own certificate.

The best option is to submit the proper paperwork and setup fee to a "certification authority," which is charged with reviewing a website's credentials (actually the credentials of the business or person operating the website). Only after the authority is satisfied that the applicant is a legitimate business operation and the genuine owner of the website in question will a digital certificate be issued.

Note: *Certification authority licensing is a mish mash. In the U.S., the various States have licensing authority, which may be exercised by, for example, the Secretary of State in one instance and the Department of Commerce in another. In other countries, a governmental agency generally issues and oversees certification authority licensing. As an example, in Malaysia it's the Malaysian Communications and Multimedia Commission.*

A certification authority (also known as a "certificate signer," or a "signer authority") will award you a digital certificate (after performing a background check) in return for a fee. There are numerous certification authorities eager to provide e-commerce sites with a digital certificate for your server. Once the certificate (which is in the form of a data file) is obtained, it must be installed on your server to activate the server's SSL features.

If a website wants to be able to conduct secure transactions but doesn't have a digital certificate, it can use someone else's server and certificate. The least expensive method is to use a hosting company's certificate. But this option has a drawback — you will need to use one of your hosting service's domain names in any secure URL. Thus, when a customer clicks a button to send his or her credit card information the customer's web browser will indicate a different URL, e.g. https://www.xyzhostingservice.com/www.yourcompany.com/order.htmlname/oderform.htm instead of your URL, e.g. https://www.yourcompany.com/orderform.html.

This may cause confusion on the customer's end, and that confusion could prevent the customer from completing the transaction.

Note: *Most web-hosting services require that an e-commerce website have a digital certificate before they will allow the website to use their services. This is a good thing — digital certificates not only allow a website to accept credit card orders securely, but also it helps to keep hackers at bay.*

Two of the largest certification authorities are VeriSign and Thawte (which is owned by VeriSign), but there are also other reliable certification authorities including Entrust, Equifax, Globalsign, and RSA Security, to name a few.

For those readers residing in the EU, the European Certification Authority Forum (ECAF) maintains a list of European Supervisory Authorities (www.eema.org/sa_matrix.asp), which may be useful when searching for a Certification Authority.

To wrap up, digital certificates are widely used by e-commerce websites, especially to provide security for credit card transactions online. If a website has a digital certificate, a customer can verify that the website displayed on their computer screen actually is what it appears to be, e.g. *your* website, not some imposter masquerading as your website in order to intercept your web visitors' communications.

Note: *For more information on digital certificates, download the free SSL Guide, which can be found on Verisign's website (www.verisign.com) and the technical introduction to digital certificates offered by Microsoft (it can be found at http://www.microsoft.com/technet/treeview/default.asp?url=/technet/security/prodtech/cyrpto/Certs.asp).*

✪ MERCHANT ACCOUNT

We've addressed how you can receive credit card information securely through the use of SSL and digital certificates, but the actual processing of the credit card requires that you have a "Merchant Account." A merchant account is a business account at a financial institution that functions as a clearing account for credit card transactions.

While there are many different payment methods, most e-commerce sites will want the ability to accept credit card payments from customers. There are two ways to process credit card payments: offline or online. Both require a merchant account and credit card terminal.

Although many people equate a merchant account with a checking account, setting up a merchant account is a bit more complicated. A plethora of businesses, in addition to traditional financial institutions, are eager to set you up with a merchant account. Thus, there is an enormous variety in the deals offered. Prior to making a final decision as to what company you will use for your merchant account, do your homework — learn about the process, talk to others who have existing accounts regarding their experience with their provider.

Offline Order Processing

If you have a brick-and-mortar business, it is likely that you already have an existing merchant account as well as a credit card terminal. Thus, you initially may want to continue with your current set-up for your new e-commerce site. Of course, that means only offline credit card processing, which would work like this: A form would be included on your website so that after a customer types in the billing and shipping information, the

Figure 9. A typical credit card processing process.

information is relayed to you through encrypted email. You then process the information manually using your existing credit card terminal.

Note: *Some customers may want to pay via their Debit/ATM card. Processing this type of payment is basically the same as processing a credit card except that the order amount is deducted from the customer's checking account. Websites that provide this type of payment option can usually forego the need for check processing. However, before making a final decision on this matter, check with your Merchant Account Provider for details on how they might handle Debit/ATM card processing and obtain their pricing information for such service.*

Real-Time Credit Card Authorization

Does your current e-commerce software allow integration with real-time credit card authorization systems? Do you process credit card transactions prior to product fulfillment? If the answer to both of these questions is yes, then you might need real-time authorization capability. (A real-time credit card authorization account doesn't require you to lease or to purchase equipment or to install software on your computer.)

If you want your customers' credit cards processed instantly 24 hours a day, 7 days a week, and the funds deposited into your business checking account within 48 - 72 hours, open a real-time credit card processing account after your merchant account is approved. Contact a real-time Internet processing company such as AuthorizeNet.com or Cybersource.com.

With a real-time credit card authorization account, authorization occurs at the time of sale; the processing network receives the authorization information from your merchant terminal and checks the databases of the financial institution that issues the card for available balance and reduces the available balance (but no money changes hands). This authorized transaction is then stored in a local database called a "batch." Settlement occurs once per day for any authorizations that have accrued for that business day. When a settlement or auto-settlement has been executed, the transaction, minus any merchant fees, will be approved for transfer to your bank account and the funds will arrive in fewer than three business days. (Settled batches, or closed batches, are stored for later access.)

There are specific requirements that an e-commerce site must meet before it can open a real-time credit card processing account. Advantage Communications Enterprises, a well-known web design and marketing firm in Kalamazoo, Michigan has provided a list of items needed to establish a real-time credit card processing account, they include:

* A shopping cart.
* Software or CGI scripts used to store products and orders.

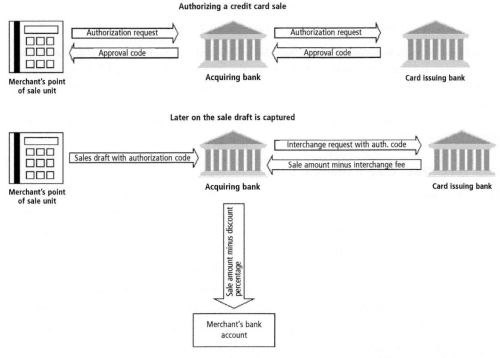

Figure 10. Steps in transaction processing when there is a website has established a real-time credit card processing account. Compare with typical credit card transaction depicted in Fig. 9. *Graphic courtesy of Advantage Communications Enterprises of Kalamazoo, Michigan.*

* Hosting for Storefront.
* An Internet connected web server.
* Business banking account.
* Internet-ready merchant account.
* A bank or merchant processor who has access to an Internet connected processing network (to enable high-speed/real-time authorization).
* A gateway to the "processing network."
* A high-speed provider with access to processing network. (There are banks and merchants that provide gateway services, and there are gateways that provide merchant services.)

For more information about real-time credit card processing, visit Advantagecommunications.com.

❂ CHECK PROCESSING

The more payment options you can provide to your customers, the more competitive your website. Some of your customers will prefer to pay by check, but accepting a check online can be problematic. Purchasing products/services by check negates the instantaneous nature of e-commerce — if you wait for a check to clear before shipping. To speed things along, some businesses will ship on the receipt of a check, at least until they are burned once too often by a bounced check.

When you process a check online you aren't required to have a signed instrument from the customer, all that's required is the information that's on the customer's regular check. There are a number of ways to accept checks online; some allow you to provide speedy order fulfillment, others require that a prudent e-commerce operator wait several days to ensure the check clears before shipping the order.

Let's first examine the do-it-yourself method. This method requires that you have a program to print out a hard copy of your customers' checks before depositing them into your bank account. (The ensuing process is the same as a traditional business that accepts customers' checks, i.e. to prevent fraud, you shouldn't ship the merchandise until the check clears.) If you are interested in going this route, check out vendors such as CheckMAN (www.checkman.com), Intell-A-Check (www.icheck.com), and Vcheck (http://softwaresolutions.net/vcheck).

To help you better understand these check processing programs, let's take a quick look at how the Intell-A-Check 6.0 application suite facilitates the processing of checks. When your customers pay by check, they provide information about their checking account by filling out a form on the website. Intell-A-Check uses this information to automatically create a check or automated clearinghouse transfer that can be deposited immediately into your bank account and immediately credited against a customer's account. You don't need to worry about not being paid because Equifax, the leading provider of consumer information in the U.S., guarantees Intell-A-Check checks. Another benefit of Intell-A-Check — the customer and the website avoid credit card fees.

Note: *Most check programs require that you purchase specific "check paper" for printing the customers' checks. But, in many cases, these programs do not require that you use any special, expensive magnetic-type ink for printing the checks — generally printing the checks on your Inkjet or laser printer is satisfactory. This is because most banks now use optical, rather than magnetic, devices to process checks. However, to be safe, contact your local bank and make sure they will be able to accept checks that you print from your computer before investing in check acceptance software.*

Now let's look at the method that will allow speedy shipment of ordered products. This method requires that you use a transaction service, which will verify that the information on the online check is complete. These transaction companies charge a setup-up fee in addition to a per-check and/or a percentage fee. Many of these services will, for an extra fee, also guarantee the check. The extra fee may be well worth it since the transaction service must reimburse you if the check doesn't clear. Of course, there are specific conditions that must be met for this scenario to play out.

Caveats:

* Most of the transaction companies in business today only process U.S. checks.

* If you decide to use an online check transaction service, be sure that the service can perform check verification in real-time.

* There is not much involved in processing a check online, so don't sign with a service that changes a high transaction fee, and don't pay a discount (percentage) rate on check transactions — unless the checks are guaranteed.

Finally, be aware of fund holds on check transactions, and know when funds will be available.

Note: Some readers may also want to investigate the feasibility of using a Secure Gateway Companies' Virtual Terminal. Some virtual terminals can accept check transactions (e.g. AuthorizeNet) by running a check verification against a national database of bad check writers, thereby reducing the incidence of returned checks. Some Secure Gateway Companies' that provide virtual terminals will even re-submit returned checks automatically.

✪ OTHER PAYMENT OPTIONS

Due to the requirements for establishing a merchant's account, many small e-commerce operators will need to find another method for accepting credit card payments. If you fall into this category, don't despair. There are still means by which you can sell your merchandise online. Two of the most widely used methods are online escrow services and email payment services.

Online Escrow Services

In the e-commerce world, trust is viewed as the intermediary element that will determine if businesses and users are willing to embrace the online economy. Online escrow services have been identified as one of the key business infrastructures necessary to help provide that trust and confidence between buyers and sellers.

When credit card fraud occurs, the e-merchant loses his goods, is charged for the costs, and must pay the issuing bank a charge-back fee. On the other hand, online consumers face the risk that they may not receive the goods, the goods that arrived are not as described, or the goods are damaged.

When you contract with an online escrow provider, the escrow service acts as an impartial trusted third party that facilitates buying and selling by providing both the seller and buyer with trust, security, and convenience. These middlemen hold your customer's payment (whether via check, money order, or credit card) in trust, awaiting confirmation that the goods are as expected. Through the escrow service, the buyer and seller agree in advance as to how the goods and funds will be exchanged, along with a return policy.

Here's how the typical escrow service works: The buyer pays the total purchase price to an escrow provider, which holds the customer's money in trust. The seller ships the merchandise directly to the buyer. If the buyer accepts the merchandise, the escrow service pays the seller. Otherwise, the buyer returns the merchandise to the seller (in its original condition) and receives a refund from the escrow service.

It's that simple — of course the escrow service charges a fee for its services. Some of the escrow services you might want to check out include Escrow.com, Escrow Online (www.escrowonline.org), Canada Escrow Online (www.escrowonline.ca), Secure-Commerce.com (also offers a multi-currency escrow service), and Cash-Escrow (http://secure.cash-escrow.net), which is a European escrow provider.

Be careful, though. In the U.S., the Federal Trade Commission (FTC) has recently begun pursuing fraudulent online escrow services. Premier-Escrow.com is one such service that, according to the FTC, set up a fake escrow firm and then directed buyers and sellers of merchandise to use that firm. And although a website would ship merchandise to the buyer, the site never receive payment from the escrow service and buyers would make payments to the fake service, but never receive their merchandise. The FTC says that it has also shut down two other alleged fraudulent escrow services and that 53 separate actions have been taken by various states concerning the same type of fraud scheme.

Here are some suggestions of steps you can take to protect yourself from being a victim of a fraudulent escrow service:

* Be wary if a buyer is insistent upon using a specific escrow service.
* If you are asked to deal through a service that you are unfamiliar with, call the company — this may help you to determine whether it sounds legitimate.
* Check out the service's website. Many fake escrow services set up websites that are intended to mirror real escrow services. Are there grammatical errors and/or other simple mistakes? Is there odd wording, especially where the fake escrow service's name has been substituted in place of the real escrow service?
* Check the Whois registry (www.networksolutions.com/en_US/whois/index.jhtml) to see if the online registry information fits with the corporate information on the site. If there are discrepancies, check them out.

✳ Check with your local Better Business Bureau.

If you have any doubts about a specific escrow service, don't do business with that company.

Some other steps you can take to ensure you are using a legitimate service is to check with your bank to see if it offers an online escrow service or can recommend one. Or bypass the escrow services altogether and use Western Union's BidPay (www.bid-pay.com). The buyer purchases a money order online using a credit or debit card, Bid-Pay sends a confirmation email to the seller within minutes with information on when your money order is scheduled to be sent. According to the BidPay website, "there is absolutely no fee for sellers unless you choose a payout option other than a money order (i.e. a cheque in British Pounds)."

Email Payment Services

Email payment services can be used as either your primary, or as a secondary payment processor. The most popular of the email payment services is probably PayPal (at least in the U.S.). PayPal makes sending and collecting money easy. PayPal integrates seamlessly with existing financial networks, allowing anyone to send money from their credit card or bank account. Thus, with PayPal, any website can accept credit card payments from all of its customers. For more information, visit the website at www.pay-pal.com.

If you live in the U.K. you might want to check out the popular NoChex.com. Anyone with an email address and a U.K. bank debit card can take advantage of this easy-to-use payment service.

For an email payment service with a global perspective check out Xcompte.com. Its services are available in multiple languages and numerous currencies.

✪ CONCLUSION

The design and pre-build details — domain name, digital certificates to ensure data is kept secure, and payment acceptance — are among the first steps you must take when extending an e-commerce site to the Web. But there are still a number of steps to take before you have a viable e-commerce website. The path you take to open your doors to the online public may differ from the path taken by your competitors, friends, and colleagues. The following chapters will guide you through some of those diverging paths.

Chapter 4

Server Hardware

If only foresight were as good as our hindsight!

—Anonymous

Behind the friendly face of many websites is a sophisticated system of servers, hardware and software. Servers are the backbone of any good website. A server is a computer on "steroids" with a very fast permanent connection to the Internet and subsystems that protect against power outages, hackers, and system crashes. (The combination of hardware, operating system, and server software constitute a "server.")

The raw performance of web servers, application servers, HTTP servers, and database servers is the key to customer satisfaction and future scalability. Slow response times are frustrating to customers and can impact the flow of data through the transaction-processing pipeline. This affects other components of your back office, including network connections and, in the end, reflects negatively on your bottom line. Nevertheless, for e-commerce sites that deal with thousands of customers at any given moment while also dispensing hundreds of thousands of unique page views per day, achieving peak performance and allowing for future growth is a somewhat mysterious art.

Unless you have a specific reason for running your own servers, outsource to a web-hosting service. You can do this through a co-location arrangement, an enterprise contract (leasing servers, applications and technical service), or rental of server space (virtual hosting). (See Chapter 14 for a complete discussion on hosting solutions.) Whether hosting the site yourself or outsourcing to a hosting company, it is advisable to learn about the necessary hardware requirements.

Let's begin by considering what is needed to set-up a small website that doesn't use intensive graphics and is destined to serve a limited number of simultaneous visitors.

✪ THE SMALL WEBSITE'S SERVER SPECIFICATIONS

For many small entrepreneurs and brick-and-mortar businesses, the main barrier to building their own website is the cost. The good news is that the hardware necessary for hosting your own website is no longer expensive or difficult to set up.

Note: Before you can understand the interrelations of the components needed to build a website, you must understand the basic terms and workings of a computer. The capabilities of your CPU, bus systems, power supplies, hardware and software are all part of the foundation upon which the website is built. If you are well versed in this technology read on. However, if you feel you need a refresher course on the inner workings of a computer, go to Appendix A: Computer Basics, where you can find a brief explanation of the various components that make up a computer.

Server Needs

The first step in determining your server needs is to determine your hardware requirements. In this section we discuss the bare-bones server specifications for a website that will be hosted on premises by a business with a limited budget. However, it is recommended that you build a web server with components that are in excess of the bare-bones recommendations.

A normal server, whether a web server or otherwise, consists of a computer (the hardware), an operating system (usually some flavor of Unix or Windows), and one or more pieces of software (the applications). The software can be web server software, email server software, commerce server software, etc.

For security reasons, your web server should be exclusive, i.e., stand-alone. If the web server is tied to your business' internal network, install another computer to house a firewall; position it between your internal network and your web server.

Memory is important. While as little as 64 MB of RAM is sufficient for a small static web server that does not house a database, always install as much memory as your budget will allow. Also if you anticipate more than moderate traffic from the get-go, expect to use a database to drive your site while means you must load your server with as much memory as it can hold. The same holds true for websites that plan to use a lot of graphics and/or sound.

Consider how much hard drive space your software and web pages will need. Again, if your website is a static brochureware site, you will most likely only be running an operating system, web server software, and traffic analysis software on your server, thus a 1 GB hard drive will be sufficient. For other website layouts, it is suggested that you install the largest drive your budget will allow.

When it comes to processors, the general rule of thumb is to go for the fastest processor your budget will allow. However, if you have a limited budget, a web server running older technology, such as a 233 MHz Pentium with MMX, should work just fine for a static website. (Don't use anything slower — doing so will limit the number of simultaneous connections your web server can handle, and it will take a longer time to load each page.) To handle more than moderate traffic, or more processor inten-sive web design (e.g. numerous graphics, database, interactive content), choose a later Pentium model and load it with as much memory as possible.

All web servers should also be equipped with an uninterruptable power supply (UPS) and some kind of backup, which can be as simple as floppy disks, a Zip drive (an inexpensive external drive), or a read/write CD drive.

To sum it up, it is possible to design a static brochureware website with a com-puter running a 233 MHz Pentium CPU with MMX, 64 MB of RAM and a 1 GB EIDE drive, *if* that is all your budget will allow. However, to grow, your website will eventu-ally need a computer on "steroids." So, while working within the constraints of your budget, go for the fastest processor, the most memory the computer can take, the largest SCSI drive with a RAID set-up, backup protection such as a read/write CD or tape backup system, and attach everything to a robust UPS system.

✪ SERVER SPECIFICATIONS FOR A ROBUST WEBSITE

The face of large e-commerce websites is deceptively simple. The friendly web pages of bookseller Amazon.com or PC retailer Dell Online (previously called Dell.com) hide the fact that huge, pulsating server farms with sophisticated load balancing and fault tolerance are quickly brokering the customers' web clicks through a maze of database accesses, cached data, and network routes.

Deciding upon the server hardware to run your website will probably engender a hard fought battle among your entire technical team —designers, programmers, and IT staff. The final server hardware configuration will be based on a "best guess" in that no one will be able to reliably predict the data traffic load that the hardware will ulti-mately carry. Keep in mind that as your website grows your server configuration will need to adapt not only to your evolving business, but also to whatever new technology comes along.

In the server world there is a saying: "You can never have too much capacity." In other words, the demand for disk space, memory, and available processing power seems to max out faster than anyone can predict. Web servers and other applications, such as database servers, can quickly expand beyond the limits of any single platform.

Server Needs

A suggested configuration for a robust website's server includes a Dual Pentium III 933 MHz or better; 512 MB RAM or better; Dual SCSI (10,000 RPM) drives; and Raid 5 configuration.

Note: The Dell e-commerce site consists of 75-plus Dell PowerEdge servers, each with one or two processors and anywhere from 256 Mbps to 2 Gbps of RAM. Because anywhere from 3000 to 4000 simultaneous shoppers visit the online store during peak hours, Dell also crafted solutions that can maintain the state of customer transactions in the event of system failures.

✪ THE TERMINOLOGY

Whether going it alone or using a hosting service, you will need to have a basic understanding of the technology used. Here are some of more common terms you might encounter during your research.

Processor Architecture

"Processor Architecture" pertains to the overall organizational structure of the computer processor unit (CPU), commonly referred to as the "processor" or "microprocessor." The main elements of any processor architecture are the selection and behavior of the structural elements and the selected collaborations that form larger subsystems that guide the workings of the entire processor. (For more detailed information on how microprocessors work visit www.howstuffworks.com/microprocessor.htm.)

The microprocessor industry is highly competitive. Consider many factors when selecting your processor architecture including performance, scalability, open/proprietary architectures, on-chip functions, advanced functionality, software availability and, of course, cost.

The Athlon (K7) is a Pentium III-class CPU from AMD with clock speeds ranging from 500MHz to 650MHz. Using a 200MHz system bus, the chip contains the MMX multimedia instructions and an enhanced version of AMD's 3DNow 3-D instruction set. The Athlon plugs into a slot, known as Slot A, which is similar to the elongated slot used by Pentium IIs and IIIs.

The Hewlett Packard PA-8000 is based upon the older PA 2.0 architecture. This 64-bit processor with a superscalar architecture can execute four instructions per cycle with its two integer ALUs (arithmetic logic units), two shift/merge units, two floating-point units, two divide/square root units, and two load/store units.

The IBM POWER4+ microprocessor is a "server on a chip" that contains two one-gigahertz-plus processors, a high-bandwidth system switch, a large memory cache and I/O. IBM's POWER family of microprocessors is among the most widely used in the industry and can be found in Nintendo game consoles, Apple computers, and some of the world's most powerful supercomputers and storage systems.

The Intel Xeon is a Pentium CPU chip designed for server and high-end workstation use. The Xeon plugs into Slot 2 on the motherboard and its L2 cache runs at the same speed as the CPU. Xeon introduced the System Management Bus (SMBus) interface, which includes a Processor Information ROM (PIROM) that contains data about the processor and an empty EEPROM that can be used by manufacturers to track their own information such as usage and service information. Xeon chips can address 64GB of memory.

The MIPS Technologies R10000 Microprocessor is a four-way superscalar architecture capable of executing four instructions per cycle, which are then appended to one of three instruction queues (integer, floating-point, or address). Each queue can perform dynamic scheduling of instructions. Although instructions are executed in order, they don't need to be, which allows the R10000 to maintain up to 32 active instructions at a time that are in the process of being executed. It has five independently operating execution units (2 integer ALUs, 2 floating-point units, and a load store unit for generating addresses).

The Sun Technologies' UltraSPARC (Scalable Processor ARChitecture) is an open reduced instruction set computer (RISC) specification to which anyone can build compatible chips and in which a microprocessor is designed for very efficient handling of a small set of instructions. The UltraSPARC architecture scales very well, ranging from low power notebooks and portables to multi-million dollar Cray scientific research supercomputers. The UltraSPARC based systems dominate the UNIX server market and provide the processing power behind many of today's robust web-hosting services and Internet Service Providers (ISPs).

Clusters

One way to surpass the limitations of a single server is clustering. Clusters are essentially multiple servers that are used as a single unified resource through the use of software, switches and routers. The inter-connection of multiple servers sharing resources results in greater availability due to the ability of the other systems in the cluster to assume the workload of a failing resource.

Maintaining access to data is a key element of the concept of high availability and if your server is running but can't reach critical data, your web-based business can come to a standstill. Clusters depend on: strong data sharing models, systems that can co-exist at different software release levels, the ability to dynamically vary systems off-line, and robust recovery models that are able to keep everything up and running. A cluster also must be able to perform load balancing. This allows processing to be evenly distributed among the various machines that comprise the cluster. In extreme cases, the software will automatically exclude a failed server from the cluster, allowing the working servers to take up the slack.

To the network's devices and your administrator, the cluster appears as a single server. This streamlines your network management efforts.

Load Balancing Switches and Routers

Other load bearing solutions include load balancing switches and routers that distribute traffic to a group of servers, sharing the load among them.

The first versions of load balancing routers were designed specifically to support Internet traffic, especially web servers. They used a round-robin algorithm, distributing requests to each server in sequence. Newer load balancing router algorithms provide a more even load distribution across a group of web servers. They react to the traffic coming through the load balancer and distribute the traffic according to the load on each server. Some balancers can check to see if a system request is for data that is residing in a server's cache, thus making it easier for that server to respond quickly. They also provide monitoring of the actual load level on each server in real time, allowing each server's load to be kept perfectly balanced.

In a server farm configuration, the servers are usually identical computers that are large enough to handle only a fraction of a website's total traffic. The number of servers needed varies. Some web operators opt for a few large servers and others choose several smaller, less expensive servers. With either option, the load-balancing algorithm can be a simple round robin, or may be more sophisticated, taking into account each server's current load. A server farm configuration provides high availability since multiple machines can take on additional processing in the event of the failure of any single server. This type of server configuration is cheaper and easier to implement than a high availability cluster since the servers don't need to be aware of each other. Only the load-balancing component is aware of all of the servers. In the event of a failure, there is no complex failover process. Instead, the system just stops sending requests to the failed server and routes traffic to the remaining servers. In this way, you get both availability and scalability.

Web server load balancing produces a cost-effective way for websites to respond to growth spurts without the need to replace equipment every few months with a new, higher-capacity system.

Rack Units

A rack unit is a vertical shelving system to mount servers. Racks can be freestanding or else bolted into the floor or wall. Racks generally hold rackmountable computers that are 19 inches wide and have faceplates that allow the computer to be affixed to the rack's frame via screws. Two or more rackmount servers can comprise a high-availability groupware (sharing data across a distributed system) and web-server system. Mounting servers in racks helps with space problems and allows for ease of management when it comes to the formulation and assignment of security measures and redundant power supplies.

Generally each "layer" of a rack holds one 19-inch rackmount computer, but smaller computers from companies such as Crystal Group, Swemco and APPRO allow for up to four or more PCs to share a shelf on a rack.

Multiple-CPU Servers

A multiple-CPU unit is a good choice for websites that demand high-end computing environments. It is a viable option used to satisfy the server consolidation requirements of very large websites that offer, in addition to products and e-commerce options, services such as free email, chat rooms, streaming video, etc.

The choice of a multiple-CPU unit should take into consideration the need for the flexibility in mixing and matching components in a rack. Choose multiple-CPU units if you are unsure of exactly what your needs are now or what they might be in the future.

Another advantage of multiple-CPU servers is that redundant power supplies are among the basic requirements (not an added feature) for an enterprise class system such as the 8-CPU Intel Servers. Consider, however, that these multiple-CPU units can cost in excess of $20,000.

Server Cabinet

A server cabinet is a metal cabinet designed to house rack-mounted servers (some also house tower configured systems). A good server cabinet will usually have a slotted front door, a perforated steel rear door and a perforated top panel to assure maximum airflow. There is room for fans or blowers to be added, which are usually necessary. The side panels usually lift off for easy and quick component accessibility.

✪ SAMPLE WEBSITE SERVER CONFIGURATIONS

Let's examine some sample real life configurations that might be found running a robust website.

Simple Website

You can build a website around an inexpensive server configuration as discussed previously, but your website would be limited in the number of simultaneous visitors it could handle. A better choice would be a simple, relatively inexpensive web-server configuration that could reliably run a website that serves, on average, around 500,000 pages per month, such as:

* Silicon Graphics Inc. Challenge S with a 180 MHz IP22 MIPS R5000 CPU and 128 MB RAM, and file systems consisting of (1) 3GB SCSI (system) and (2) 4GB differential SCSI2 (web pages).
* The software might include the Unix IRIX 6.2 OS, Apache 1.3.1 web server software, and Perl 5.003 CGI Development Platform.
* Tape Backup 8mm DLT.
* ISP connection via either a symmetrical DSL line (i.e. a high-speed line that provides the same speed for both upstream and downstream data loads) or a fractional T-1/E-1 line.
* Everything would be plugged into a robust UPS.

Complex Website

Now, let's look at a couple of configurations that might reliably be used to run a more complex website:

Example 1

* Dell Pentium III 500 MHz equipped with 256 MB PC100 RAM, Dual 9GB Ultra-Wide SCSI drives, the necessary number of network cards, and Raid V Backup.
* The software might include Windows NT Server 4.0/ Windows 2000 Enterprise, Internet Information Server 4.0 (Service Pack 5.0), MDAC 2.1x, and a ColdFusion 4.0 Professional Server.
* ISP connection via a T-1/E-1 line.
* Everything plugged into a robust, long life UPS.

Example 2

* Dual Pentium III 450MHz processor equipped with 1024 MB of RAM, and running Linux, RedHat-6.2 install, Linux Kernel 2.2.20, Apache 1.3.9, and Perl 5.
* Backup would be handled by a 70GB RAID-5 array.
* ISP connection via dual fiber-optic T-1/E-1.
* Everything plugged into a robust, long life UPS.

Enterprise Website

Now let's move up to a complex website running not only a web server but also other servers including a database server. This type of site might run a very expensive, but necessary, high-end configuration such as:

Example 1

* One Sun E450, 250 MB RAM, three 400 MHz UltraSPARC II processors with 4 MB cache, eight 9 GB (10,000 RPM) hard drives with backup plan, running Veritas' Volume Manager RAID solution, Solaris operating system, Apache web server, F-Secure ssh 2.0, and WebTrends Enterprise Reporting Server for Solaris.

* One Sun E250, 128 MB RAM with one 400MHz UltraSPARC II Processor with 2 MB cache, five 9 GB (10,000 RPM) hard drives with backup plan, running Veritas' Volume Manager RAID solution, Solaris operating system, Apache web server, F-Secure ssh 2.0, and WebTrends Enterprise Reporting Server for Solaris.

* One Sun E250, 128 MB RAM with two 400MHz UltraSPARC II Processors with 2 MB cache, five 9 GB (10,000 RPM) hard drives with backup plan, running Veritas Volume Manager RAID solution, Oracle Database Server Enterprise Server for 2x400 MHz CPUs.

* One Compaq ProLiant 1850R, 128 MB RAM, One 500 MHz Pentium III Processor with two 512 K caches, 4x1-inch Hot Plug Drive Cage and 2x1 Hot Plug Drive Cage, three 9 GB (10,000 RPM) Hard Drives with backup plan, SMART Array 3200 Controller running Microsoft Windows NT Server 4.0, Microsoft IIS 4.0, Symantic PCAnywhere 8.0 Server.

* ISP connection via either a dual fiber-optic T-1/E-1 or a T-3 line.

* Everything must have a redundant power supply and be hooked up with multiple network cards.

Example 2

* Clustered Sun E450s with multiple processors and 1 GB RAM with the Solaris operating system and Apache web server.

* A database server with the same configuration except running Oracle 8.1 in place of Apache.

* An e-commerce server running two Dell 450s with 1 GB RAM and the Windows NT operating system and Microsoft IIS 4.0.

* A staging server running two Dell 450s with 1 GB RAM.

* A terabyte (TB) of external disk packs and a RAID 5 disk array.

* ISP connective via either a dual fiber-optic T-1/E-1 or a T-3 line.

* Everything should be powered by redundant power supplies and have multiple network cards to support High Availability.

❂ CONCLUSION

A web server can be a simple piece of hardware running fewer than 500 lines of code that can take web browser commands, retrieve files based on those commands, and send the files down the wire to the browser. But many e-commerce operations need a more robust server configuration; some will even need a full-blown enterprise-level web server set-up. This means the hardware and software can get more involved, but even then the basics are still simple.

If you want to host your own servers, the best advice is to create a balanced plan with each segment working in partnership with each other. An application can run so fast on Pentium-class hardware that the application can saturate the network before the database or the server become saturated. This can cause trouble. Everything must work as a partnership. For instance, while database access techniques and web application performance are important, the size of web pages (particularly when graphics and ad banners are included) often have the biggest performance impact. This is because the slowest thing is getting graphic objects to the users' browsers.

Those of you that are hosting your own website, remember that the key is to design everything as a system. You can't design your web server without regard to your database and other applications.

Chapter 5

Redundancy

Caution, though often wasted, is a good risk to take.

—Josh Billings

Redundancy is a safety measure where you install multiple units of all of your critical hardware devices so that if a unit fails the system can continue uninterrupted. Redundancy of two, three, or more times may be used to support the operations of a website including its switches, routers, and other components.

✪ BACKUP

Server crashes and hard drive failures are inevitable due to equipment failure, lightning, power outages, simple age or defect-related failures, bugs, hackers, viruses, and, of course, human error. The most basic method of saving your data is to "backup." All backup solutions have one thing in common — they involve copying data from your hard drive(s) to a second media, from which you can restore your data in the event that your hard drive(s) or your server fails.

If you create and maintain your own website, keep local and up-to-date backups of all files. At a minimum, a daily backup should be made of all the data on your server, so that if the data is erased or modified in error, much of it can be restored. Backup should be a ritual and scripted so it becomes a part of your daily routine.

Institute a backup plan that allows you to get your website up and running in seconds. Meeting your website's backup needs requires:

* A secure place to keep your backed-up files that is accessible at all times.
* A combination of software and hardware to handle the backups.

* A tape backup on your server, allowing you to perform your own data backups remotely.
* An online backup service (if you don't use a tape backup) where you can transfer your data electronically to a secure location. Or you can use the service to schedule automatic backups of selected files.

The general idea in all cases is to backup fully all of your data to tape every week or two. Every other day perform a modified backup on a new set of tapes each day. At the end of each week, take the set of tapes from the previous week and move them to off-site storage.

There are two kinds of modified backups, each with significant differences:

Differential Method (one full backup + several differential backups): With this method, you back up *only* the data that has been modified since the last *full* backup. This is done when you set the backup software so that it will leave a file's "archive flag" *unchanged* after the back up is completed. This method gives you redundancy — the original full backup and the most current differential backup. Using the full backup and the latest differential backup, one can safely restore an entire hard drive. This method requires a lot of tape space, however.

Incremental Method (one full backup + several incremental backups): This method backs up the files that have been modified since the last backup, *whether* full or incremental. Setting the backup software to *clear* the file's "archive flag" after it is backed up does this. This requires minimum tape space but may require several tape backup sets to find a lost file. To restore an entire hard drive you must restore the full backup and then restore each incremental backup in the cycle.

Get the biggest and fastest backup system you can afford.

When mapping out a backup plan, consider all the costs including hardware, software, and staff time. It usually takes at least an hour to back up a single server. Manually backing up several servers weekly, semi-weekly, or daily represents a considerable amount of man-hours.

One good reason for using a reliable web-hosting service is that most provide planned scheduled backup services. A good web-hosting service will have a backup system that is designed to prevent the loss of data that a website might experience due to server crashes, hard drive problems, and hackers.

✪ UNINTERRUPTABLE POWER SUPPLY

An uninterruptable power supply (UPS) is a device with a built-in battery that sits between the power supply and your server(s). It protects your equipment from power outages, brownouts, sags, surges, bad harmonics, etc., which can adversely affect the performance of the system. UPSs are available in numerous configurations. A UPS that can protect a single web server will cost around $250. If you have a network of servers, UPS costs can run into thousands of dollars.

There are two types of website — the ones that have had a power problem and the ones that will have a power problem. A UPS will be one of the most important pieces of equipment you will install to help ensure the reliable operation of your website.

Standby Power Supply

There are many UPSs with varying capabilities. A "standby power supply" or "offline UPS" is not a true UPS. It won't protect your web server. This standby power supply's power comes directly from the power line, until the power fails, then a battery-powered inverter takes over. The time required for the inverter to start providing electricity to your server varies greatly. Some servers might tolerate a standby power supply, but don't chance it.

Hybrid UPS

A "hybrid UPS" is a device that uses a ferroresonant transformer to maintain a constant output voltage between the power source and your server, protecting against line noise. It can maintain output relying on its battery (a secondary power source) for a limited period of time. If power is not reinstated, a total outage occurs. It is questionable whether this type of device can actually respond when needed without an accompanying interruption in power. There is some debate as to whether a UPS's ferroresonant transformer will interact with the ferroresonant transformers in your equipment, producing unexpected results. The hybrid UPS system is comparatively cheap, but, if you choose it, be sure to test it thoroughly with all of your equipment before going on-line.

True UPS

All e-commerce operators should use "true" UPS systems. While these systems are more expensive to purchase and maintain than the others we have examined thus far, this system continuously operates from an inverter with no switchover time and offers good protection from power problems.

A true UPS has internal batteries and can absorb small power surges. It continues to provide power during line sags, negates noisy power sources, and provides power for a set length of time during a power loss. It provides continuous power independ-

HOW TO ASSESS YOUR UPS REQUIREMENTS

How much will you pay for the reliability of a true UPS? To find the answer, decide the minimum and maximum amount of time you want to keep your equipment running after the power goes down. To assess the amount of time your system might need supplemental power support—which can be considerable with tape back-up and CD burners—add up the power requirements of the hardware you are running by using the equipment manuals (not the rating plates on the equipment). If power is stated in Watts, then multiply that figure by 1.5 (some experts say 2.0) to get the VA rating, which is the maximum number of Volts Amps the piece of equipment can deliver.

The energy delivering capacity of a power system is measured in Volt-Amps (VA). In the U.S., a standard electrical system outlet for the home is rated at 120 Volts of Alternating Current (120 VAC) and 15 amps (depending upon the thickness of the wire), and thus has a rated VA capacity of 120V x 15A = 1800VA. Exceed that limit and you'll trip a circuit breaker or cause a fire.

Similarly, for a UPS, an expensive, heavy-duty 300VA device has a lot more capacity than an inexpensive 100VA device. If you use a 100VA rated device anyway and exceed its rating you could blow its fuse or it will not work correctly when called upon during a power failure.

In high school you may have been taught that Watts = Volts x Amps, so you might jump to the conclusion that VA = Watts. It isn't quite that simple.

Transformer- and capacitor-operated devices have a spec called the "power factor," which is a number between zero and one. The power factor represents the ratio of the energy used by the system to the energy required by the system to make it operate properly. Light bulbs, space heaters, toaster ovens, etc., use all of the raw energy put into them and so these devices have a power factor of one. Devices based upon more delicate and complicated electronic components do not necessarily use all their power rating, all the time, and thus have a power

ent of the outside power supply. The minimum support you want from any UPS system is 30 minutes, enough to survive short outages and other power inconsistencies. Keep in mind though, as you design your server system that a true UPS generates quite a bit of heat; so don't put it in a closed space.

Included in the "true" UPS category are systems that provide:

* Automatic shutdown and restart of your website's equipment during long power outages.
* Monitoring and logging of the status of the power supply.
* Display of the voltage/current draw of the equipment and the voltage currently on the line.
* Alarms on certain error conditions.
* Short circuit protection.

How to Rate UPSs

When researching your UPS requirements and the various brands and configurations factor of less than one, generally 0.5 or 0.6. Thus, a computer may need 160VA to run correctly but it actually only uses 80 Watts so the power factor is 80 / 160 = 0.5. Another way to find the VA is to remember that VA = Watts / Power Factor.

This also means that Watts = VA x Power Factor. You might be tempted to look at your computer, see the Watt rating of 80 Watts, then run to your friendly neighborhood computer store and buy an 80 Watt UPS. Bad idea! Things are not that simple, since Watts = VA x Power Factor, so 80 Watts = 160VA x 0.5 for our imaginary computer. This means that you need a 160VA unit to protect and supply power for an 80 Watt computer. Unfortunately many UPS manufacturers use a generous power factor of 1.0 when they advertise the ratings of their VA devices and are thus listing the rating in Watts instead of in VA capacity. They figure that "Watts" is a more familiar electrical term to non-techies than VA and, besides, a power factor of 1.0 gives the largest value for the Watt rating, so 80VA x 1.0 = 80 Watts. This causes the unsuspecting consumer to buy an underpowered UPS. An "80 Watt" UPS may really be an 80VA unit that can actually only handle a 40 Watt computer! To protect a 160 Watt computer, you would have to buy a 320VA UPS.

And even if you think you've "figured out" what the right size is for your UPS, be sure to add another 20% capacity for good measure.

What this all comes down to is that, if you want your website to be up and running during a blackout, then you will need a very robust UPS. Your UPS will rely on its battery (DC) to AC converter, which means an expenditure of power. Just to give you an idea at what you are looking at, the author will hazard a guess that a 1250 VA UPS could probably operate during a blackout for around 5 hours and a 2000 VA UPS could operate during a blackout about 8 hours.

Before purchasing a UPS, be sure to ask the vendor about guarantees; if the guarantees don't fit your needs, find another vendor.

available, make certain that the UPS vendor offers a support and/or maintenance contract. If not, go to another vendor. There are many options you may wish to consider before purchasing your UPS. A manual bypass switch is helpful so that when the UPS is out of operation, power can pass through it to your web server. You also should know how close the AC output of the UPS is to a sine wave.

An inverter is an electronic device that converts a battery's DC output to AC through a switching process, producing a "synthesized" AC, which can be charted as a waveform on an oscilloscope or graph paper. Inverters produce two types of waveforms: The so-called "modified sine wave" and the "true sine wave."

True sine waves, or sinusoidal signals, are the most common waves that exist. They're called sine waves because they have the same shape as the graph of the sine function used in trigonometry. Sine waves look the way they do because they are produced by rotating electrical machines such as generators and, indeed, a sine wave's intensity (amplitude) at any given instant can be represented by a point on a wheel rotating at a uniform speed since waves are perfectly "balanced" over successive time intervals.

On the other hand, a "modified sine wave" is not actually a sine wave, but a stepped wave, which is the kind of wave a pendulum produces, and is not as smooth as a sine wave. But "mod sine" inverters cost half the price of sine wave inverters, thus lowering the cost of a UPS. However, they can cause electrical noise on a circuit, and digital clocks and timing circuits can be confused or even be occasionally damaged.

Sine wave inverters in a good UPS deliver true sine wave AC output power with high efficiencies from storage batteries. They have high surge ability and low idle current draw. Because of the pure sine wave, the expense is greater than a modified sine wave inverter.

If your UPS does not output a pure sinusoidal waveform, do not put a surge protector between the UPS and the server since a surge protector can mistake the non-sine waveform as a power surge and send it to ground; that will damage the UPS. Some experts think that most computers use a switching-type power supply that only draws power at or near the peak of the waveform, therefore the shape of the input power waveform is not important. It is the authors' opinion that it is worth spending a little more for a UPS with pure sinusoidal output especially for a UPS that must continually provide a waveform to the computer.

Check what useful operational information the UPS itself provides via displays, etc., such as the power or percentage load the unit is drawing, the battery level and power quality. Most UPSs use lead-acid batteries with a life span of only a few years but no battery memory. Therefore they should be run "dry" as few times as possible.

Maintenance

As your UPS ages, its battery life will become shorter. Be vigilant in monitoring the active support time. When your website has no one accessing it (or you can take it off the Internet for a few minutes) and you've completely backed up you hard drive(s), test your UPS and its failure modes. Simulate a power outage by throwing the circuit breaker that has the UPS on it (don't pull the plug from the wall) to check the UPS. If you don't have your UPS-protected website on an isolated circuit, you could install a Ground Fault Interrupter (GFI) socket (GFI sockets are the electrical switches with a red and a black button you have in your house or office).

Power problems are inevitable and beyond your control. Therefore, a UPS is one of the most crucial items you can purchase to keep your website up and ready for action.

✪ RAID

RAID, which is an acronym for "redundant array of independent (or inexpensive) disks," should be a part of all but the smallest websites. By purchasing a good UPS you've protected your site against power problems. Now you need to protect your website against data problems and drive failure. That's where RAID comes in. Although some may feel that a RAID system is too expensive, give it serious thought.

A RAID system links the capacity of two or more hard drives that are then viewed as a single large virtual drive by the RAID management software. By doing this it is possible to improve data storage reliability and thereby achieve fault tolerance.

A basic RAID system includes RAID functionality built into a controller and two or more hard drives. While RAID software can implement RAID in a server without a special drive controller, the efficiency and performance leaves much to be desired. Don't use it.

RAID can be found in many different configurations and in just as many price ranges:

* A floor-standing cabinet.
* A complete system in one full-size drive bay.
* A self-contained system with its own redundant power supplies, etc.
* In RAID Levels 3 and 5, drives can be hot swapped (you can change drives without shutting down the server) and the RAID controller and reconstruction software will automatically rebuild any lost data.

RAID Levels

Although there are many different RAID levels, the most common ones used for website operation are listed below.

RAID-0 divides each file into blocks and distributes these among multiple disks in a process called "disk striping." This provides high performance since more than one disk is read and/or written to simultaneously. A file can now be input with one revolution of four disks as opposed to four revolutions of one disk. Unfortunately RAID-0 doesn't have the one key feature you expect from a RAID subsystem: Data redundancy, hence no fault tolerance. When you read that a system supports RAID-0 it really means that although the system has disk striping, it isn't actually RAID at all. RAID 0 is used for high performance situations such as video editing and is generally not used for web servers, unless a high performance database also must be put online.

User's view of storage is one large disk
A B C D E F G H I

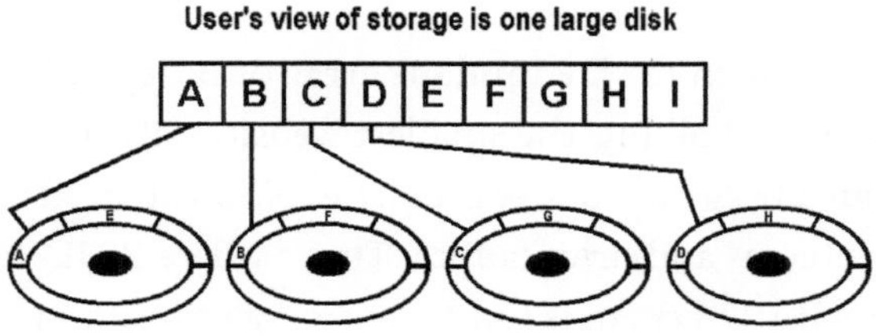

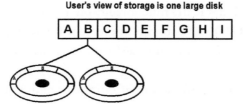

Figure 11. RAID-0 — Data is divided and striped across multiple drives. RAID-0 is typically used to increase a system's performance, but this type of RAID offers no data protection.

User's view of storage is one large disk
A B C D E F G H I

Figure 12. RAID-1 — Data is completely copied or "mirrored" onto a second disk. This type of RAID offers good data protection, although this is expensive for a large website.

RAID-1 is the easiest and, for a small website, can be one of the least expensive ways to protect your website's data from a hard drive failure. With RAID 1, as the data is written it is simultaneously copied or mirrored onto a second disk which is connected to a common disk controller and, *voila*! You have data redundancy. RAID-1 is considered to be the most common, secure, and reliable form of RAID. However, as your need for data storage increases, your costs can become considerable. At such a point you would look for another storage method such as RAID-3 or RAID-5.

Before going further into RAID technology we should explain parity data in the form of a type of Error Correcting Code (ECC), which avoids the cost of duplicating disk drives in their entirety. There is a method of transmitting binary data where an extra bit (the "parity bit") is added to each group of bits. If parity is to be odd then the extra or parity bit is assigned either a one or zero so the total number of ones in the character will be odd. If the parity is even, the parity bit is assigned a value so that the total number of ones in the character is even. In this way errors can be detected.

The author knows that this explanation probably appears as clear as mud to the reader, but stay with me and hopefully it will become clearer.

By using RAID with parity, when a drive fails the ECCs and binary value of the striped bytes or sectors can be used to recover data from a failed drive by comparing data on the still-functioning drives to the parity data that sits on a special parity data drive. The RAID system then can re-create the data on the failed drive.

This is similar to how one solves a missing variable in an equation. For example, 3+5=8, where "3" is a bit on one drive, "5" is a bit on another drive, and "8" is the data's parity information stored on a third drive. If the drive storing "3" fails, you could recalculate it by solving for X in the equation X+5=8, so X=3.

Or, to state it in very simple terms, parity-based systems calculate the data in two drives and store the result on a third drive. Those results can be later used to reconstruct what was on the other two drives. Now we can continue with our RAID discussion.

RAID-3 stripes data across multiple disks one byte at a time. Parity is also calculated bit-by-bit and stored on an extra "parity drive." All drives have synchronized rotation. When a drive fails, data is rebuilt transparently in the background from the remaining functioning drives as the system continues to operate.

RAID-5 is the most popular high-end RAID technique used today. RAID 5 stripes data at the sector or block level across a minimum of 3 drives. It also provides stripe

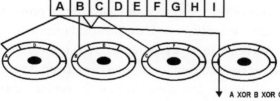

Figure 13. RAID-3 — Data is striped across multiple disks one byte at a time. Parity is also calculated byte-by-byte and stored on an extra "parity drive." All drives have synchronized rotation. Although RAID-3 offers a good data redundancy system, RAID-5 is better.

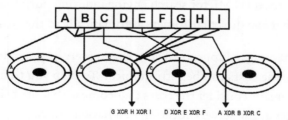

Figure 14. RAID-5 — Data is striped across multiple drives in large, sector-sized blocks. Drives spin independently. Parity information is striped along with the data. RAID-5 is the most popular RAID configuration.

error correction information by striping it along with the data evenly over the drive set. This results in excellent performance and good fault tolerance but it still lags behind the performance found with RAID-1 disk mirroring. Most of the high-end, pre-configured RAID set-ups are RAID-5. A RAID-5 system with preconfigured drives, RAID-5 software, adapter cards, and the necessary cables, is easy to purchase and to set up.

RAID-10 is really just a combination of RAID-1 and RAID-0, i.e., mirroring and disk striping. While expensive, RAID-10 is a reliable, comprehensive RAID set-up and should be considered when operating a large, full-blown e-commerce website.

Some kind of RAID storage is necessary for your website and there are certainly all types and levels available. To learn more about RAID systems, read Richard Grigonis' book, *Fault Resilient PCs* (Miller Freeman).

✪ MIRRORING

Mirroring of a website is the creation of multiple websites that are exact duplicates of an existing website. Only high-traffic sites should consider adopting a mirroring model.

A website can use mirrored sites in many ways:

* To improve the download speed to your customer by providing more than one server in more than one location with the identical information. Therefore, when traffic becomes too heavy for one geographical site (or one server) to handle, your system can hand off the additional traffic to the mirrored site.
* Live standby servers where data contained on the primary servers are seamlessly mirrored on the standby servers. In the event of an unplanned outage, such an infrastructure can keep a website running efficiently, including both hardware and software.
* When a website uses a system that generates dynamic HTML, it can create a mirror of the site with fixed HTML for search engine indexing purposes, along with special scripting that can direct customers automatically to the "real site."

Since you are creating a duplicate of your site in every way, including the software and the hardware, the costs can be astronomical. To help you to decide whether to implement a mirror site, consider the costs that come with a site being down because of a natural disaster at the servers' location or a hacker attack.

✪ CONCLUSION

Despite the most compelling design, and innovative programming, there are many reasons that a website can fail. One of the keys to minimizing site failure is redundancy. The power and data redundancy discussed in this chapter are a good start to ensure that your website is always available, but other types of redundancy also should be considered when building a website. For example, using industry-grade equipment and ensuring your site has more than one route to the Internet.

Chapter 6

Connectivity

Take care of the minutes and the hours will take care of themselves.

—Anonymous

"Connectivity" is a buzzword used in Internet circles to refer to the ability of computers to link to networks, and therefore to the Internet. This chapter deals with the bandwidth aspect of connectivity, i.e., the amount of data that can be transmitted in a fixed amount of time. Although this chapter does not provide a complete education on connectivity and bandwidth, it does provide the information you will need to make an intelligent connectivity decision.

The first step is to estimate the amount of bandwidth you will need to feed data from your web server to your customers.

Bandwidth (in the Internet context) is the number of bits per unit time that can be carried across a communications line. The basic unit of bandwidth measurement is bits per second (bps), but most commonly you'll see it expressed Kbps or Mbps, which stand for kilo-bits per second and mega-bits per second, respectively. The rate at which your connection transfers data is measured in the same bits per second. This bandwidth determines how fast data is transmitted to and from your web server and also how many requests can be serviced simultaneously. Even with a robust server, if you do not have sufficient bandwidth for the number of customers coming to your website, delays or failures will occur.

Select a way of delivering bandwidth that is scalable to meet your website's future needs while at the same time limiting additional costs and frustration. For example, although a small website can get by with a business-class DSL connection, this solution does not have much scalability. If you can afford it, you should choose to go with at least a burstable or fractional T-1/E-1 line. A large or enterprise website may want to go with a T-3 connection.

Since your website should be accessible to everyone on the Web, and you hope to have many visitors to your site, plan for lots of bandwidth. A website offering streaming video, lots of audio, or one that receives a large amount of traffic would need a T-1/E-1 connection all to itself.

In addition to the number of simultaneous customers, think about the speed at which data is sent to your customers. This is determined by connection speed and data size. It should take fewer than five seconds to send a page of type. Even with the additional bandwidth demands of graphics, audio, or video, a page should load in fewer than 30 seconds.

Once you've determined the amount of bandwidth needed for your web server (whether self-hosted or outsourced to a web-hosting service), consider whether your website will offer other services that require increased bandwidth such as email, chat rooms, streaming audio or video. If so, be sure to include enough bandwidth to cover those services.

If running a self-hosted site, you should understand that the Internet connection comes through a router. A network interface card connects your web server to the router — a high-performance network card will prevent a bottleneck between your Internet connection and your web server.

Whatever size bandwidth you end up using, know that most Internet service providers (ISPs) or web-hosting services will take care of most of the details of getting the line installed. But you should still understand that your local telephone company is needed to provide the piece of the connection (the "local loop") that brings the Internet to your door. Then, if your web-hosting service has a point of presence (POP) co-located with your telephone company's central office (practically all do), you'll be able to connect with them using just a local loop. Otherwise you'll have to rent a dedicated line to wherever the "on ramp" happens to be.

It is up to the web host (whether or not an ISP) to ensure that their POPs are capable of delivering the bandwidth and the response times that their customers need. Ultimately, these providers must be able to expand their POP as they increase their customer base and add new services.

Now let's look closer at the high-bandwidth services available.

✪ DSL

DSL (Digital Subscriber Line) technology brings high-bandwidth transmission to homes and small businesses over ordinary copper telephone lines. "xDSL" refers to different variations of DSL (e.g. ADSL, HDSL, and RADSL) that provide high-bandwidth transmission of up to 10 Mbps over your local telephone company's ordinary copper wire lines. Installation of Asymmetric DSL (ADSL), a popular version of xDSL, appeared in 1998 and soon exploded to offer service throughout the United States and elsewhere. In many areas it has replaced ISDN (discussed later in this chapter).

DSL works by placing special line conditioning and transmission equipment on both ends of an ordinary copper line from your location to the local telephone company's central office. Because this connection uses a much broader range of frequencies to transmit digital data than a standard analog phone line, higher speeds are possible. Your data is then routed from the telephone company's central office to your web host/ISP over high-speed trunk lines, then out to the Internet.

DSL offers several advantages over an ISDN connection. Even a "lite" ADSL connection offers from 384 Kbps to 1.5 Mbps of Internet bandwidth as opposed to 128 Kbps with ISDN, i.e., DSL is at least six times faster. Full-rate ADSL can move 8 Mbps, thus having the bandwidth of several 1.5 Mbps T-1/E-1s.

The Different DSL Technologies

Technology	Downstream Rate	Upstream Rate
ADSL (Asymmetric DSL)	1.5 to 8 Mbps	640 to1.5 Kbps
ADSL Lite, G.Lite	384 Kbps to1.5 Mbps	128 to 640 Kbps
CDSL	1 Mbps	128 Kbps
EtherLoop (symmetrical)	125 Kbps to 6 Mbps	Same
G.Lite	384 Kbps to 1.5 Mbps	128 to 384 Kbps
HDSL (High bit-rate DSL)	768 Kbps to 2.3 Mbps	Same
HDSL2	1.5 Mbps	Same
Hotwire ReachDSL	512 Kbps 320 Kbps	Same Same
IDSL (ISDN DSL)	144 Kbps	Same
RADSL (Rate-Adaptive DSL)	7 to 8 Mbps	1 to 1.5 Mbps
SDSL (Symmetric DSL)	160 Kbps to 2 Mbps	Same
SHDSL	Many bit rates: 192 Kbps to 2.3 Mbps	Same

Table 2. DSL comes in a number of variations. If possible find a DSL service that provides the same speed both upstream and downstream. Many times such a service is marketed as "business DSL."

DSL is hundreds of dollars per month cheaper than a burstable or fractional T-1/E-1 connection. A DSL connection gives you acceptable bandwidth to your web server at a fraction of the cost. But DSL is not very scalable.

Since most DSL technologies require installation of a signal splitter there is an up-front installation expense of about $100.00. You also will need to install a DSL router that is connected to your web server's network hub, which will cost around $75.00. Use only a high-speed "business class" DSL service (i.e. a symmetrical DSL technology that offers the same speed upstream and downstream) to host a web server. Your monthly costs for business-class DSL service should be between $75.00 and $300.00.

✪ ISDN

In some areas, the only "high-speed" service offered will be ISDN (Integrated Services Digital Network). ISDN is an access technology that uses digital transmission with a bandwidth of up to 128 Kbps over your local telephone company's ordinary copper wire lines. However, unlike your local telephone service, which transmits analog signals, ISDN can, over these same wires, transmit multiple digital signals simultaneously. There are two ISDN user interface standards: Basic Rate Interface (BRI) and Primary Rate Interface (PRI).

* ISDN BRI is defined as consisting of two 64 Kbps B-channels and one low-speed 16 Kbps D-channel (used only for signaling).
* ISDN PRI standards vary according to the geographic region the service serves. In the U.S. and Japan, ISDN services based on the PRI interface are defined as 23 B-channels plus a D-channel (all operating at 64 Kbps), yielding a total of 1.536 Mbps). In Europe, however, ISDN PRI is defined as 30 B-channels and one D-channel yielding a total of 1.984 Mbps.

Wherever situated, the slower D channel is always the last channel, (e.g. channel 24 in the U.S. and channel 31 in Europe), and it serves the purpose of call signaling, call setup, requesting network services, and routing of data over B-channels.

A terminal adapter (TA) is the piece of equipment you install at the end of an ISDN line in your home or business -- it can be an ISDN phone, a fax machine, or an ISDN "modem."

There are two kinds of ISDN TAs, internal and external. TAs can look like modems, like computer bus cards, or like interface cards for PBXs or routers. Many ISDN phones and modems have analog jacks on the back that allow you to connect ordinary phones, fax machines, and other analog, non-ISDN devices.

External ISDN TAs are easier to install (they plug into your PC's serial or USB port) but won't give you maximum performance. Some, such as the Motorola Bitsurfr Pro or the 3COM Courier, actually look like modems and are often called "ISDN modems."

The problem with ISDN is that it's not scalable — you get 128 Kbps (a little over twice the speed of a 56k modem) and that's it. Data is routed from your ISDN modem to your web host/ISP over the same copper wiring that you may have previously used for ordinary analog phone service, then out to the Internet.

ISDN is available from the local telephone company in most areas in the United States and Europe for around $150 for installation, $300 for the external adapter/modem, and a $50 to $100 monthly line fee. But use ISDN only if there isn't another, faster option available, e.g. DSL.

⊙ T-1/E-1

A T-1 (E-1 in Europe) line's data-carrying capacity is 1.5 Mbps (E-1 offers 2.048 Mbps) within 24 channels (E-1 offers 31 available channels), each channel having a 64 Kbps data transfer capacity. For most small businesses a full T-1/E-1 line may be overkill since such service is expensive. However, a fractional T-1/E-1 line or burstable T-1/E-1 may suit even a small website's budget.

A fractional T-1/E-1 is a T-1/E-1 line that is channelized or partitioned, and is referred to as a "fractional configuration." If a business does not need a full T-1/E-1 line it can lease any portion of the 24/31 64 Kbps channels, with the transmission method and rate of transfer remaining the same. This service is most preferable for a website that expects its traffic to be higher than 1 Mbps 50% of the time.

A burstable T-1/E-1 is a cost-effective Internet access solution for websites to receive direct, reliable, high-speed Internet connectivity. You only pay for the bandwidth that is used rather than for the total size of the circuit and bandwidth, much of which is often unused.

Access to the Internet is set at a minimum of 128 Kbps with burstability to the full T-1/E-1 capacity of 1.5/2.048 Mbps when necessary. Although a monthly minimum is based on sustained usage of 128 Kbps, burstable T-1/E-1 service always provides the availability of the full T-1/E-1 bandwidth. An unshared, point-to-point, full T-1/E-1 line can cost anywhere from $800 to $1500 per month, but a burstable or fractional T-1/E-1 is less — how much less will be based on the service you need.

For example, here's how the cost for a burstable T-1/E-l might be calculated: Usage is usually monitored every day with SNMP (a network monitoring and control pro-tocol) to create an end-of-month usage report. Rates are then averaged based on the middle 90% of reported usage for the monthly basis. Reports should be included in your monthly billing.

Any T-1/E-1 installation (including burstable and fractional service) will come with a set-up fee of approximately $5000 that includes:

* All the local telephone company loop charges and carrier fees.
* Full 1.54/2.048 Mbps availability.
* Cisco router or other certified network equipment.
* T-1/E-1 CSU/DSU.
* All T-1/E-1 installation charges.
* T-1/E-1 set-up fees.
* As many IP addresses as can be justified.

✪ WEB HOST'S/ISP'S CONNECTIVITY

You also need to consider the size of the pipe of your web host/ISP. When searching for the best ISP for your self-hosted web server or a web-hosting service to host your web server, ask pointed questions about their connectivity strength. If you expect your website to receive say, a million hits a month, then find a service that can provide you with sufficient data feed, that means the provider must have T-3 and/or optical carrier lines.

Here is a quick tutorial to help you understand what these different data feeds provide.

T-3/E-3

A T-3/E-3's (also known as DS3) data carrying capacity is about 45 Mbps with 672 channels, each channel having a 64 Kbps transfer capacity. This type of service is very expensive and overkill for all but the largest, most popular websites (and websites offering multiple video streams). Many Internet Service Providers (ISPs) use T-3/E-3 lines as their connection to the Internet backbone.

Even if you feel that your website might need a T-3/E-3 data stream, first consider using inverse multiplexing (the reverse of ordinary multiplexing, which combines multiple signals into a single signal). Generally, inverse multiplexing across even eight T-1/E-1 lines is less expensive than going the T-3/E-3 route.

Inverse multiplexing speeds up data transmission by dividing a data stream into multiple concurrent streams. Those streams are transmitted at the same time across separate T-1/E-1 channels and are then reconstructed at the other end back into the original data stream. It is a technique commonly used where data in a high-speed local area network (LAN) flows back and forth into a wide area network (WAN) across the "bottleneck" of a slower line such as a T-1/E-1. By using multiple T-1/E-1 lines, the data stream can be load-balanced across all of the lines at the same time.

Optical Carrier (OC)

The term "Optical Carrier" describes fiber optic networks that conform to the Synchronous Optical Network standard for connecting fiber-optic transmission systems (commonly known as the "SONET standard"). Fiber optics is a technology that uses gloss or plastic fibers or threads to transmit data. A fiber optic cable consists of a bundle of fibers or threads, each of which is capable of transmitting messages modulated onto light waves.

The SONET standard also defines a hierarchy of interface rates that allow data streams at different rates to be multiplexed. SONET establishes OC leaves from 51.85 Mbps to 2.488 Gbps. The standard OC levels include:

* Optical Carrier Level 1 (OC-1) offers data speeds up to 51.85 Mbps.
* Optical Carrier Level 3 (OC-3) is actually an optical carrier with three 51.84 Mbps multiplexed OC-1 circuit streams on an underlying SONET/SDH circuit and thus offers data speeds of up to 155.52 Mbps.
* Optical Carrier Level 3-c (OC-3c) is actually an OC-3 circuit with the three OC-1 lines concatenated into a 155.52 Mbps circuit (used in ATM transmission).
* Optical Carrier Level 12 (OC-12) offers data speeds of up to 622.08 Mbps. An OC-12 line can handle over 400 times more data than a T-1/E-1 line. Some Internet Service Providers, especially those that also offer web-hosting services, often provide OC-12 connectivity to the Internet backbone.
* Optical Carrier Level 24 (OC-24) offers data speeds of up to 1244 Mbps or 1.244 Gbps. Canada is at the forefront in deployment of OC-24 service. Both Le Groupe Videotron Ltee (the second largest cable service provider in Canada) and Videotron operate and own more than 2484 miles of installed fiber-optic cable and fiber-optic infrastructure throughout the province of Quebec. Among Videotron's product mix is transmission of its multiplexed signals using OC-24 service.
* Optical Carrier Level 48 (OC-48) offers data speeds of up to 2488 Mbps or 2.488 Gbps. The Abilene, which connects regional gigaPops to form the Internet2 network, runs over a OC-48 fiber optic cable.
* Optical Carrier Level 192 (OC-192) is the newest SONET interface rate. It offers data speeds of up to 9.952 Gbps.

✪ CONCLUSION

All e-commerce operators need a good understanding of connectivity. The size of your "pipe" to the Internet has a profound affect on your customers' surfing experience. Thus, when searching for a web-hosting service for your website, consider only those that can offer you at least a T-3 line to the Internet.

When considering your own connectivity needs (whether for your internal network or a self-hosted website), neither go too small, nor too large.

It is impossible to over-emphasis the importance of connectivity. Take the time and effort to choose the right Internet connection for your website's needs. Do your homework. Your pipe to the Internet must be able to provide a continuous data stream so your website can always perform at its peak, while at the same time, giving you the most "bang for the buck."

Chapter 7

Security

Even the lion must defend himself against gnats.

—Anonymous

The Internet's openness makes it the perfect platform for e-commerce — it offers an inexpensive mass communication media and an economy of scale for low-cost distribution. However, the lack of security of web-based transactions and the ease with which the privacy of online communications can be violated are e-commerce's main stumbling blocks. Internet's very openness means that all communication traveling over it is inherently difficult to secure. To make matters worse, hacking is an epidemic that is on the rise.

Ira Winkler, president of the Internet Security Advisors Group in Severna Park, Md., and author of "Corporate Espionage" (Prima Publishing, 1999) succinctly states the average e-commerce business's security dilemma: "To a hacker, you're just an IP address. You get hit because you let yourself be an easy mark."

Here are some eye-opening figures to contemplate: A study by Gartner Inc. indicates that 50 percent of all small to midsize enterprises were hacked in 2003, with almost 60 percent of those not even knowing they had been hacked. According to the Computer Emergency Response Team (better known as "CERT," www.cert.org), a total of 82,094 incidents were reported in 2002. But, as Fig. 16 shows, incidents are rapidly increasing — there were 76,404 reported incidents in just the first half of 2003.

Don't be an easy mark. Recognize and appreciate that you are building your business in a domain that is, at least in principle, fraught with danger. Thus all e-commerce businesses must take the necessary steps to ensure that adequate levels of security

CERT's Record of Incidents Reported 1988-2003

1988-1989

Year	1988	1989
Incidents	6	132

1990-1999

Year	1990	1991	1992	1993	1994	1995	1996	1997	1998	1999
Incidents	252	406	773	1,334	2,340	2,412	2,573	2,134	3,734	9,859

2000-2003

Year	2000	2001	2002	1Q-2Q 2003
Incidents	21,756	52,658	82,094	76,404

Total incidents reported (1988-2Q 2003): **258,867**

Table 3. This CERT record of incidents is only the tip of the iceberg — a single incident as recorded by CERT, while reportedly involving only one site, can actually include hundreds (or even thousands) of websites. Furthermore, an incident can involve ongoing activity for long periods of time. *Graphic courtesy of CERT.*

are in place. This means, at minimum, firewalls to control the flow of data, monitoring software to protect web pages, and an encryption method to protect transactional data — consumer information, credit card numbers and your own merchant data. Supplement that with diligent oversight, which includes reporting and analyzing the security of your web business's entire infrastructure.

✪ DEVELOP A SECURITY PLAN

Although website security is an extensive subject that is well beyond the scope of this book, this section will discuss the main points of a good e-commerce security plan, as a piece of a website's overall reliability architecture. With an effective security system a website can create an environment that promotes e-commerce and private communications by establishing a climate that is safe from robbery and fraud.

Security is a subject most business executives try to avoid since they feel that discussing their business' security procedures and policies might add to the risk of an attack. However, without such a discussion, it is difficult for these same executives to be aware of the constantly evolving technology that can help a web-based business.

As you develop your security plan, always be cognizant that a secure e-commerce environment requires:

* Access control, usually managed by a firewall, which regulates the data flow.
* Authentication, which binds the identity of an individual to a specific message or transaction.
* Data privacy and integrity, which ensures that communications and transactions remain confidential, accurate and have not been modified.

While there are a number of security concerns that must be addressed by all web-based businesses, some websites' security needs will differ from others. For example:

* Some, but not all, e-commerce operations will necessitate opening up specific data resources (e.g. databases) to trusted third parties. But while doing so, the e-commerce operator must ensure that that same data isn't accessible by other, unauthorized, parties.
* Most e-commerce sites will want to allow their customers to know how much a widget attachment will cost as the order is being entered, but that same customer should not be allowed to see the cost of another customer's order.
* Many sites will want to allow their customers to be able to view, copy, and print the status of all of their outstanding orders via the business's website, but at the same time block those customers from copying the entire database(s) in which the information is stored.

Many of today's systems do not readily handle this level of security, but that is just what is needed to support a good e-commerce website. Although we won't visit specialized security needs as discussed in the preceding bullet list (only you know what security systems will best protect those assets), we will look at what it takes, generally, to protect a website, its contents, and customers.

✪ LAYER YOUR SECURITY

A website needs a total security architecture, i.e. security that exists in a number of layers — from the web server, to the applications, to the database, and to the extensions to other subsystems. Most brick-and-mortar businesses will have some kind of security program already installed, but, in all probability, it is not up to date. And if any part of the security architecture is not working as planned, then your whole security set-up is vulnerable.

Unfortunately, some e-commerce businesses mistakenly use security tools, techniques, and strategies that cannot withstand sophisticated security attacks. Their websites are open invitations to hackers to break into their networks (many times without the business owner even being aware that there was a break in). Once inside hackers can steal money (from the business's merchant account), products (by placing fraudulent orders), and sensitive information including customer identities (by obtaining access to the business's databases). Once inside your network, cybercriminals also can commit crimes against other merchants by using the hacked network as their launching pad. But you can minimize your risk, by implementing some of the many security solutions available that provide efficient data sharing without compromising confidentiality, availability, and integrity of the data.

Limit outside access. This is the first line of defense for any website. Some methods for accomplishing this are:

* Firewalls.
* User account security.
* Software security.
* Additional protection for sensitive data.

Protect your web server. The second line of defense is optimizing your web server so that it can resist most hacker attacks. For instance you must install antivirus software.

Implement monitoring and analysis solutions. The next line of defense is putting into place routine monitoring and, if your budget allows, analysis systems so that you know who and what is connecting to your systems, and interacting with your servers. At minimum, you should install log analysis software (see Chapter 8) that will allow you to monitor system logs and network traffic for anomalies. Simple log analysis software allows you to identify attempted security breaches and possibly to track their origin.

Better yet, if your budget allows, get a good monitoring system that enables you to analyze internal and external firewall activity and identify attempted security breaches. There are high-end security monitoring solutions available that can detect and resolve security vulnerabilities in your web-based business' systems either on demand or at regularly scheduled intervals. Look at www.webtrends.com and www.pgp.com to give you an idea of what is available in the way of security suites.

Encryption. Then follow through with the next line of defense — encryption. This refers to a system that uses encoding algorithms to construct an overall mechanism for sharing sensitive data. Encryption is the security cornerstone for most e-commerce sites.

Use a Web-hosting Service. The last line of defense is to consider using a web-hosting service — a good hosting service will have the financial means to provide the resources for truly effective website security. These businesses provide a professional staff that has the skill and wherewithal to keep abreast of the latest news and technology updates as well as the ability to implement fixes and upgrades at a moment's notice.

✪ COMMON WEBSITE SECURITY MEASURES

Now let's examine in detail some of the more common security measures you can take to protect your website from attack by hackers, vandals or other trespassers.

Routers

Be sure that your router is appropriately configured. A router is an electronic device or, in some cases, software in a computer, that forwards traffic (i.e. data) between networks. The forwarding is based on network layer information and routing tables. A router is designed to route packets efficiently and reliably, but not securely, thus although it is a layer in your security package, a router should not be used alone as a method for implementing a security policy.

DENIAL OF SERVICE ATTACK

One of the most common types of security attack is what is called a "denial-of-service" attack, i.e. an attacker or attackers use various means to prevent legitimate website users from accessing a site. The problem is that the latest forms of denial of service attacks are difficult to counter since they can happen at any given time and, at the moment, there is no solution to the problem. A website operator must address each attack as it occurs.

Examples include:

- The "Ping of Death" uses a test packet larger than allowed, which can cause a system crash or problems with network programs running on the targeted computer. Ping is the acronym for Packet Internet Groper, a program that tests a TCP network by sending a packet with an echo request to a designated host's IP address and then waiting for a reply.

- Attempts to "flood" a network, thereby preventing legitimate network traffic, and attempts to disrupt connections between two machines, thereby preventing access to a service.

- The Mail Bomb that sends a flood of mail to a mail server, sometimes overloading it, thus causing legitimate users not only to be denied service but also to lose mail that had been sent to their mailbox.

- Host System Hogging, where a program is actually run on the system of the website that is under attack causing a domino effect. It ties up the system's CPUs, the operating system crashes, the site goes down and, finally, customers can't access the site.

Not all service outages are outright denial-of-service attacks. For example, some denial of service attacks might be controlled so as only to cause degradation in network traffic. This, in turn, slows down the website, but does not block it completely. Your website could also be the victim of other types of attack such as an attacker using your anonymous FTP storage area as a place to store illegal copies of commercial software, which consumes disk space, and generates more network traffic.

Routers spend all their time looking at the destination addresses of the packets passing through them and deciding which route to send them on. A router is like a trip planner. Tell the planner where you want to go and it directs you via the shortest route.

If you are a brick-and-mortar business running a network of computers, you probably already have routers for your network, but know that the considerations are different when designing router configuration for your new website. Mainly, you must configure your router so that it can help to control per-user access from the public side (the Internet) to your web service and still protect your private network assets.

Firewalls

Install a firewall. A firewall is a device that controls the flow of communication between internal networks and external networks, such as the Internet. It controls "port-level" access to a network and a website. A "port" is like a doorway into a server. For example, your Internet request isn't just immediately sent to a server; instead there is a port number on the server that is the actual destination of a request. It's like sending a message to the server on Main Street in Elmville, New York, USA but forgetting to provide the specific "building" number. For example, http://www.yourname.com by default uses port 80 (the building number) on the www.yourname.com server. If a request arrives at the wrong server or the wrong port, the service handling requests on that port will ignore the request.

When looking at firewall solutions, you should understand that all firewalls are technically software — some just come with their own server. You also need to know that firewall solutions run the gamut from software designed to protect a personal PC to products that provide a wide range of security, flexibility, and protection. Costs can range from $50 or less from the personal PC solution (don't buy this type of software), to a $10,000 firewall/router solution, to a $100,000+ system that consists of firewall software, routers, and proxy servers. To ensure you get the right firewall system to fit your website's needs, hire an expert in data security (if your budget allows).

Since firewalls are an important component of network security, some established businesses take the position of "we already have a firewall therefore we are adequately protected." But, it's more complicated than that. A firewall must be correctly configured to provide effective protection. For instance, if your business already has in place a router running firewall software, that solution will probably not provide the security that a website needs — of course, the site's actual security needs depend on the website and the applications it is running.

Here are some examples of firewall configurations you might want to implement.

* Close off the possibility of unnecessary or unauthorized traffic accessing your servers.
* Configure the firewall so that only wanted traffic gets through.
* Encrypt most or all traffic between servers.
* Limit the points of access.

A properly configured firewall also can act as a filter to prevent suspicious requests from ever arriving at the server or can be configured to drop any request that tries to address a server or server port that has not been specifically enabled by the policy of the firewall. More importantly, firewalls can verify that the request matches the kind of protocol (e.g., HTTP, FTP) that is expected on a particular port.

Firewalls can be used in various places. For example:

* Between the Internet and the web server to limit the number of ports and protocols open for use by outsiders.
* Between the website and the web-based business's internal network to protect back-end servers and data by isolating public servers from the rest of the internal network, somewhat like a high fence.
* To isolate sensitive website data from other servers through a configuration that allows access only from the website application servers. This type of set-up is used to safeguard critical data such as credit card numbers.

Disable Nonessential Services

Although you should use firewalls as the first line of defense, they are only part of a good comprehensive security solution. You must also shut down all nonessential ports and services on production servers. Some of the services, like FTP, are inherently insecure because they send their password without encryption. Other services such as netstat and systat actually put forth information that can assist cybervandals with certain types of attacks on your website.

Some of the services you should disable on your website's servers include, but are not limited to:

* Mail (SMTP).
* Finger.
* Netstat, systat.
* Chargen, echo.
* FTP.
* Telnet.

* Berkeley UNIX"r" commands such as rlogin,rsh, rdist etc.
* SNMP.

User Account Security

A common method hackers use to gain access to a web server is to steal an authorized user's account. Restricting a user's access to only the needed resources limits the amount of damage hackers can do to your website. Authentication and authorization are the two best general ways to restrict access.

Authentication. This verifies that you are who you claim to be. An authentication system can use login-passwords, digital signatures, and a one-time password (single sign-on). An example of usage is the need to authenticate a user in order to login to a web server from a web browser.

Authorization. This defines what a user is allowed to do. It typically is used with an access control list service (ACLS) or a policy that restricts access to computer resources. Authorization may be attributed to users, user groups, or user profiles.

User IDs and passwords are the most common means of providing authentication services with authorization to access specific resources such as file directories, read/write permission, database access. User IDs are usually easy to figure out, since many are based on a user's name. So, make the passwords harder to guess. Still, this won't solve all of your problems because there are a number of tools and techniques available that can be used to decipher a password.

Here are some steps you can take to improve password security:

* Never transmit user IDs and passwords in the clear — use encryption techniques.
* Make the passwords a combination of mixed-case letters and numbers.
* Test your passwords by using a tool like password+ or Password Appraiser v3.20.
* Keep the number of accounts on production servers to a minimum.
* Use techniques for encrypting transmissions and certificates for authentication such as Kerberos (a security system that authenticates users but doesn't provide authorization to services or databases, although it does establish identity at log-on) and SSL for the administrator accounts.
* Never grant more access to resources than is needed. For example, if an application server running your product catalog needs to read its information from a database, it will need a database user account, but this account must not be authorized to read from other parts of the database (where credit cards are stored), or be allowed write privileges.

Data Confidentiality

Confidentiality ensures that only authorized people can view data transferred in networks or stored in databases. Protecting sensitive data like credit card numbers, inventory, etc. is a difficult problem for web-based businesses. To protect your data, you first need to identify the information that is sensitive. Once identified you can then take the necessary steps to make the data harder to retrieve; that can be done by using any or all of the following methods:

* Put the data on a separate server behind a firewall.
* Separate the data. For example, give a database its own security subsystem and user authentication process.
* Restrict the number of user accounts that can read/write the data.
* Separate write access from read access.
* Encrypt the data and control access to the encryption key.

Software Security

Internet applications often require security services such as authentication, data confidentiality, data integrity, and nonrepudiation. The major web server applications install their own security implementations on top of the operating system.

Any application that processes requests from a user is seen as a separate component in a website's architecture. Applications do handle incoming requests and therefore need to ensure that requests stay within permitted boundaries. The first step in providing some kind of security for these applications is not to trust the correctness of user input — whenever input is received it is validated. Why? Because it is possible to "piggyback" a second command onto an input request to a Unix shell by separating the two commands by an ampersand (&) or semicolon (;). Therefore, every input request should be parsed for validity or filtered for suspect content. A second step is to use separate security subsystems to control access to an application's resources where users are authorized and authenticated. This way, the application administrator, not the system administrator, controls the application's security system. As such, normally there is no one person who can control all of the information relating to a website. Also, e-commerce packages usually come with security subsystems to control their particular resources.

Nevertheless, a web-based business's security methods should be centralized to the largest degree possible. This allows you to limit accessibility to security information and eliminates the need to extend trust to many administrators. Finally, every application running on a website must consistently apply a clear security architecture (and must

MINIMAL WEB SECURITY FOR YOUR WEBSITE

All websites are targets for hackers, but these troublemakers especially like to pick on the small e-commerce site, because many of such sites don't have robust security measures in place. Fortunately, even a small web business can implement relatively inexpensive security against cybervandals. Basic security such as maintaining passwords, encrypting and password-protecting your business's data, and monitoring all website visitors is a good start. Although it is virtually impossible to make any website 100% secure, you can implement security features that so discourage the would-be hacker, that he or she takes their vandalism someplace else.

Here are some minimal web security strategies that website owners could take to help ensure their website is uninviting to cybervandals. While some of these suggestions are inexpensive to implement, others can cost a bundle. It's up to the reader to decide how proactive he or she wants to be when it comes to protecting their e-commerce business.

First, let's address the minimal security needed for a self-hosted website.

1. Install a firewall and a router. The cost for this solution can range anywhere from $1,000 to $10,000, depending on the size of your website and the type of protection you want to install.

2. Put your web pages on a CD-ROM and then have your CD-ROM drive — not your hard drive — feed the website. While it is relatively easy for some unauthorized individual to access a server's hard drive, it is very difficult to penetrate a CD-ROM drive. Note, however, that while this security method will slow down your website; it is still a good, viable option for small web businesses. Read/write CD-ROM cost less than $200 and blank disks are dirt cheap.

3. Although it may be tempting for the novice website owner, don't put website content in your server's administrative account (a file folder that holds operating keys to the network). This is the first place that hackers look.

consistently fit within your site's overall security architecture), making explicit its operational requirements.

Another step is to keep abreast of all security alerts for threats against your type of systems. Once alerted to a new technique for breaking into your type of system, counterattack with a plug and keep repeating the scenario — plug a leak, the hackers come up with a new way to break in, plug it, the hackers formulate a new break in, plug it, and on and on.

It isn't easy to make an operating system secure, but it can be done if you know where all the leaks are so you can plug them. Keep on top of your operating system's weaknesses, i.e. what is being exploited by hackers — this will allow you to keep (hopefully) one step ahead of them.

Firewall applications remove unnecessary, resource-hogging services and at the same time plug potential security holes. This is easier to do with Unix-based firewalls than with Windows-based systems since developers have easier access to the Unix code.

4. Protect all server access-control lists with a password and monitor the list regularly to ensure only authorized users are listed. If a hacker can break into the list, they will add themselves to it — allowing them to roam your website's systems at will.

5. Be pedantic about scripting. Errors in CGI scripts (the program language commonly used to create website features) are notorious as cybervandals most popular point of entry. Use precaution with JavaScripting because this program language is full of places where hackers can get in and change your website. Set up your website scripts properly. Ask your programmers how they are safeguarding the JavaScript features created for your website. Finally, if feasible, consult with a security expert on how to set up your scripts.

6. Use Secure Socket Layer (SSL). This will help to prevent hackers from detecting both passwords and credit card transactions as they travel the cyber byways. SSL is part of most e-commerce software programs, but some website owners turn SSL off to speed up their customers' web experience. Accept that although SSL may slow down your server a bit, it is necessary for the success of your website.

7. Do you allow your customers to track orders and shipments online or want to provide them with this amenity? If so, separate the customer-accessible data on a separate network from the rest of your business. The best way to do this is to isolate the customer-accessible web server with its databases from your other data assets.

If you use a web-hosting service, security steps three through seven apply. But also add to your security to-do list the following:

- Keep an ISP contact list handy in case of a breach.
- Be proactive in determining the service's security quotient. (Later in Chapter 14, we discuss how to determine if a web-hosting service offers adequate security for your website.)

Microsoft is much more protective of its code and therefore is not as cooperative with developers.

Content Security

The most commonly reported website security problem is website defacement and sabotage. In fact, cybervandal-based site defacement is an ever-present and increasing threat to website owners worldwide. To illustrate this troublesome trend, according to London security consultancy mi2g Ltd., website defacements totaled 20,371 in the first half of 2002, up 27 percent from the 16,007 recorded in the same period the year before.

Note: According to mi2g Ltd., over the last seven years, the worldwide economic damage estimate for all forms of digital attack is at between $118.8 and $145.1 billion (as of August 2003).

The complexity of many of today's websites (with their numerous pages, images, and associated features) means that manual methods such as looking at each piece of content and repairing it, as needed, are ineffective. Thus today, a website owner may want to put in place an intermediate server. Lockstep Systems Inc. (www.lockstep.com), Tripwire Inc. (www.tripwire.com), Watchguard Technologies Inc. (www.watchguard.com) and others provide software that resides on an intermediate staging server to ensure that your site's content is staged and preserved before deployment to the actual website. At configurable intervals, the staging server queries your website to compare files for differences. If an anomaly is found, the contents and/or files are captured and quarantined by the staging server, and the original content is restored from the intermediate server's files. The software also logs the incident and sends an alert to the website's administrator.

Monitoring Your Website

Finally, monitor your website's usage and take a proactive stance on security holes. To ensure a high level of security, you should:

* Monitor for break-ins. Institute a user account change report or install a sophisticated network monitoring system.
* Monitor your logs after an attack, they can tell you how the attack occurred and might even provide a clue as to the identity of the attacker.
* Run a security analysis program that can take a snapshot of your site and then analyze for potential weaknesses in your site.
* Perform security audits with outside auditors to check for potential security holes that you might have missed.
* Back up your website on a scheduled basis so that, if needed, you can recover damaged data and programs.

Security and Certification

Most web-based businesses go further than the security provided at the router and firewall level. They incorporate such features as encryption of credit card information and other personal data, digital signatures and trust of identity of network users, hosts, applications, services, and resources.

Internet trust services, including authentication, validation, and payment, are needed by websites to conduct trusted and secure electronic commerce and communications. Digital certificates provide trust of identity, which enables a website to conduct online business securely, with authentication, message privacy, and message integrity, all helping to minimize risk and win customer confidence.

As explained in detail in Chapter 3, digital certificates (DC) can be compared to a driver's license. DCs provide an electronic method by which you can prove the identity of a specific computer's owner/operator. For the driver's license, a credible organization (the DMV) assures that the driver's license is issued to the correct person. The same is true for DCs, a certification authority (CA) issues digital certificates but only after verifying the identity of the entity/person. The DC contains among other things, the name of an entity/person, the entity/person's public key, the serial number, and the signature of CA (which was signed using the CA's own private key).

Credit Card Security

With the proper precautions, online purchases are no more dangerous than credit card purchases made in the physical world. E-commerce systems keep credit card information secure by encrypting the information. Most online purchase transactions are encrypted using the Secure Sockets Layer (SSL). SSL is an internationally accepted standard for the secure transmission of data. Virtually all web browsers and web servers have standard SSL capabilities built into them. Thus nearly all web browsers and web servers communicate with each other using SSL.

As explained in Chapter 3, the SSL-driven process handles all of the security for most transactions by enabling the customer's browser to confirm the identity of the server it's dealing with, and providing an encryption system that ensures that all data is transmitted safely.

There is also a security standard called Secure Electronic Transactions (SET). SET encrypts a credit card number so that only designated banks and credit card companies can read the information. SET requires you to obtain a special certificate from your bank, and then your customers must install special software on their computer. The software is supplied by various vendors, most of whom seem to have the word "wallet" somewhere in the name. The "e-wallet" software allows your customer to input all of their purchasing information (credit card, address, shipping address, etc.) once and then move merrily through numerous websites that accept that e-wallet technology, doing "one-click" shopping and avoiding the repeated task of filling out individual websites' purchase forms.

So far, U.S. consumers have had little incentive to use the e-wallet applications that SET requires, and because the SET systems are costly and complex to set up, the SET standard, although readily accepted outside the U.S., is not widely used by U.S. web merchants and banks. The main barrier has been that each e-wallet provider has established their own unique technical specifications, so many U.S. websites have adopted a wait and see attitude with regards to SET.

Also alternative options have emerged for handling credit card transactions over the Web that are easier and cheaper than SET. The next-generation e-wallets use a new standard — the Electronic Commerce Modeling Language (ECML). This standard works with any web security software and allows e-wallets automatically to feed customer information into the payment forms of participating websites. Visa, Master-Card, and American Express, with support from America Online, CyberCash, IBM, Microsoft and Sun, as well as numerous web-based businesses have led implementation of the ECML standard.

ECML can be used with any security protocol, including SSL, and Visa and MasterCard's own version of SET. Therefore, ECML may change the landscape for e-wallet companies and websites — increasingly you will find e-wallet services offered by financial institutions and credit card companies.

Some of the more popular websites offering e-wallets include:

* www.passport.com (.NET Passport e-wallet from Microsoft).
* www.iliumsoft.com (e-wallet from Ilium Software).
* www.gator.com (e-wallet from Gator Corporation).

Now let's review how you can prevent your business and your customers from becoming victims of Internet fraud.

○ INTERNET FRAUD

How serious is Internet fraud? To quote Verisign, a company that delivers critical infrastructure services that make the Internet and telecommunications networks more intelligent, reliable, and secure, "The threat of online fraud is so pervasive that the government has begun mandating security requirements for businesses that handle financial information online." Although currently such regulations only apply to the banking community, e-commerce businesses access the financial networks for each transaction made on their website. Thus, security at the point of sale is an increasing concern for not only governments, but also for credit card associations.

These sobering figures show how prevalent Internet fraud has become:

Gartner Group estimates that fraudulent transactions make up 1.06% of total online transactions versus only .06% of offline transactions. Gartner also estimates that in 2003 alone, online transaction fraud will reach $1.8 billion.

The FBI reported that in 2002 Internet fraud complaints tripled from the year before; and, sadly, 2003 complaints are above the level reported for the same time period in 2003.

Although any e-commerce site can be at risk and a single fraud incident may be serious enough to put a merchant out of business, some websites are at greater risk for certain types of fraud than others. Some of the *higher than average risk categories* include e-commerce sites that:

* *Don't have robust security defenses* — cybercriminals can take advantage of such sites using sophisticated spidering techniques to enable them to search the Internet for websites with network vulnerabilities. Criminals then use this information to break into your network where they can steal your account access information for hijacking or merchant takeovers.
* *Are highly visible* — although you, of course, need to have a high-visibility quotient to attract customers, fraud attempts are higher for e-commerce sites that advertise heavily or are media darlings. That's because cybercriminals understand that websites experiencing high volume traffic spend less time defending against fraud.
* *Sell internationally* — it is difficult to validate the address or identity of out-of-country customers. It is even more difficult to investigate and prosecute fraudulent activity internationally.
* *Offer seasonal or special promotion merchandise* — criminals know that a website owner has limited time to spend on fraud protection measures when sales volumes are high.
* *Sell high-ticket items* — criminals fraudulently acquire items that can be resold easily.
* *Sell downloadable goods (e.g. software, music)* — the purchase of these goods doesn't require a physical address, which makes it easier for criminals to disguise a fraudulent transaction.

There are steps you can take to significantly reduce your exposure to fraud. These steps are separated into three levels: the individual transaction level, the account level (i.e. protecting access to your payment gateway account), and the network level. However, to protect your business from fraud, you must address each of these levels in an integrated manner.

Transaction Level. This is where you ensure that each transaction you process is a valid transaction. To do this you must authenticate the customer and screen order for fraud patterns.

* Take advantage of MasterCard and Visa's buyer authentication programs.
* Put in place a system that offers address verification service, card security code features, IP address checks, shipping address validation (e.g. VeriSign's Payflow service).
* Maintain a list of "bad" or fraudulent orders and check all transactions against that list *and* a list of repeat customers who have previously transacted legitimate business on your website.

✳ Don't automatically reject a transaction that seems suspicious. Instead, review such transactions to ensure that you aren't rejecting a legitimate customer.

Account Level. At this level you ensure that only authorized users have access to your payment gateway account. Also put in place a system whereby you can be alerted for suspicious account access patterns. For instance:

✳ Lock down administrative access.

✳ Change your account password on a regular basis.

✳ Monitor account level activity for suspicious patterns that could indicate merchant account takeover.

Network Level. This is where you ensure your network or "perimeter" is defended against unauthorized access. As described in this chapter's "Common Website Security Measures" section, protection at this level includes:

✳ Locking down network access.

✳ Monitoring firewall activity.

✳ Putting in place a system whereby alerts and patches are automatically checked and installed on all servers, operating systems, and applications.

✳ Investing in regularly scheduled security audits or port scans to identify network vulnerabilities.

✪ SECURITY RESOURCES

Some of the websites that offer good, reliable security information include:

✳ Cisco Systems Security Products and Technologies (www.cisco.com/warp/public/cc/so/neso/sqso/index.shtml).

✳ Microsoft Security Advisor (www.microsoft.com/security).

✳ Sun Microsystems Software and Networking Security (www.sun.com/security).

✳ The digital risk specialists, mi2g Ltd. (www.mi2g.com).

A good website to bookmark for Internet trust issues is www.verisign.com. There you will find free seminars, white papers covering trust and security issues and other resources.

Also bookmark the following websites — they can help you to keep abreast of the latest security threats to your computer systems.

✳ Computer Incident Advisory Capability (www.ciac.org/ciac/).

✳ Federal Computer Incident Response Capability (www.fedcirc.gov/).

✳ Advanced Laboratory Workstation System (www.alw.nih.gov/Security/security-docs.html).

✳ CERT (www.cert.org).

✪ CONCLUSION

If your budget allows, retain a security expert to perform a detailed review of your web-based business' internal procedures, network topology and permissions, access controls, hardware, software, and utilities that could possibly compromise your website.

Please note that even if you set up intricate levels of security, your website is never completely safe from a determined and skilled attacker. E-commerce operations are particularly hard to protect since they must be able to interact with their customers. Therefore, at the very least, build and maintain a good, state-of-the-art firewall and encrypt sensitive data, such as credit card information. And don't forget to institute an on-going program of security monitoring, maintenance, and to perform an annual security audit.

Chapter 8

Infrastructure Software

If you do not think about the future you cannot have one.

—Anonymous

If your web-based business is to cross the finish line of the "race for success" you must choose the right software. When making your software choices consider ease of integration, scalability, and robustness, i.e., does it work well with other software and hardware, can it scale as your business grows and as technology changes, and can it handle large amounts of traffic.

The first software decision to be made is which operating system to use for your e-commerce business. Next, choose your basic website software: web server, log analysis, and database. Specific e-commerce software is such a broad subject that it has its own chapter — Chapter 9. Then there is adjunct software, which many of you will want to consider — those are discussed in Chapter 10.

Note: Software can be defined at its most basic as computer instructions or data — anything that can be stored electronically can be software.

⊛ THE OPERATING SYSTEM

An operating system (OS) acts as a computer's CEO — whether it operates on a server or a desktop computer. Just as a CEO controls all of the operations of a corporation, the OS controls all of a computer's operations, such as allocating memory, accessing disk drives and calling applications, to name just a few. The software applications that provide the bulk of an OS's services run on top of the OS. The three most popular operating systems for web servers are UNIX, Linux (a freeware version of UNIX), and some flavor of Windows (NT, 2000, 2003 Server).

UNIX

UNIX is an interactive time-sharing OS that comes in many variations or "flavors." UNIX is not a single operating system; it is a family of operating systems with each vendor providing its own version of UNIX. Although these various versions of UNIX are similar, there are enough differences that you might have problems if you decided to move (or "port") your application from one version of UNIX to another.

Some examples of the various manufacturers' UNIX versions are:

IBM's AIX L5, which is an integrated UNIX operating environment that provides full interoperability and coexistence between 32- and 64-bit applications with processes that may run concurrently or cooperatively, sharing access to files, memory, and other system services. The inclusion of Internet technologies such as Java and XML parser for Java as part of the base operating system enables AIX to be Java-ready for both Java client and Java server applications.

Silicon Graphics (SGI) IRIX 6.5, a scalable 64-bit operating system based on UNIX standards for high-performance graphics and scalable server systems. The IRIX operating system allows you to take full advantage of MIPS processor-based SGI systems. IRIX 6.5 was specifically designed for users with the most demanding technical compute and visualization needs, IRIX 6.5 readily scales to tackle huge data sets, compute-intensive problems, and real-time 3D visualization enhancements with ease.

Compaq/HP Tru64 UNIX V5.1B is a 64-bit operating system offering a choice of management interfaces, including web-based operations. This system reduces the complexity of installation, setup, and management compared to other UNIX systems. It supports multiple terabytes of data, increases performance in file system, storage management, and system networking, and delivers very high integration between UNIX and Windows operating systems.

NetBSD is a free, highly portable UNIX-like operating system available for many platforms, from 64-bit AlphaServers to Macintoshes and handheld Windows CE devices. Its clean design and advanced features make it excellent in both production and research environments, and it is user-supported with complete source code. Many applications are easily available.

Santa Cruz Operation (SCO) OpenServer systems are among the most popular of operating system platforms for small- and mid-sized businesses. The latest generation of SCO OpenServer systems allows you quickly to expand your computing environment as your business's computing needs grow. SCO's integrated support for email and Internet services ensures that your business can leverage the Web to give your company the exposure it deserves and to communicate more efficiently with customers and prospects. Aside from critical business applications, SCO OpenServer Enterprise System provides various network services including file and print services for UNIX and Windows systems, email services, web services, Internet connectivity, and calendar services. Built-in support for Windows File/Print services means that a single multi-functional server platform can be utilized to run your entire operation, while the field-proven reliability of SCO OpenServer keeps you in business.

Sun Microsystems (Sun) Solaris 9 is a UNIX operating system. There are various flavors of Solaris 9 including a SPARC version and an Intel version. If you have the budget and you are self-hosting your website, the Solaris 9 operating system should be considered since it sports some hefty security features, including IPSec with Internet Key Exchange (IKE), extensible password encryption, secure LDAP, and Kerberos. The Solaris 9's strength is its enterprise-class reliability and scalability and its mature networking kernel. Solaris 9 supports hot-swapping of processors and has a management console that provides a simple, customizable utility for maintaining and monitoring Solaris 9 servers, along with wizards for installation of applications.

Linux

Linux (which is a variant of UNIX) is a free operating system and therefore can be copied and redistributed without paying a fee or royalty. It is a multi-user, multi-tasking operating system that runs on many platforms, including Intel processors. Linux interoperates well with other operating systems, including those from Apple, Microsoft and Novell. There are many Linux packages offering a full and rich set of utilities, connectivity tools, and a development environment. There is even free software available that turns Linux into a very nice web server. The main drawback is that there is no formal technical support for Linux.

Most programs written for either Linux or UNIX can be recompiled to run on the other system with little reconfiguration. Linux in most situations will run faster than UNIX on the same equipment. You can download a copy of Linux from the Internet or purchase an inexpensive CD set. You'll find Linux is a first rate operating system with capabilities beyond what you expect from more expensive products.

A crucial issue to consider when making your operating system decision is that UNIX (and Linux when configured properly) are inherently more secure than Microsoft products. Since UNIX was designed as a multi-user multi-process platform for interconnected computers, guaranteeing security has been an issue ever since the beginning. All UNIX vendors publish fixes for their software whenever a security problem is found. Microsoft is still somewhat secretive about security problems and thus there can be quite a lag time from the moment Microsoft learns about a security problem and when the public is notified about it.

Windows 2000 and Windows 2003 Server

While Windows 2000 Server is essentially an upgrade of Windows NT, Windows 2003 Server is much more. Although both Windows 2000 and 2003 include a bundled web server — Microsoft's Internet Information Server IIS — the version is different: IIS 5.0 is bundled with Windows 2000 and IIS 6.0 comes with Windows 2003.

Both operating systems offer a great array of innovative network tools and business features; however these features can cause the OS to be a challenging operating system to install and maintain, especially if you are unfamiliar with the inner workings of Microsoft operating systems.

Nonetheless, if you feel you are up to the technical challenges of Windows 2000 Server and/or Windows 2003 Server, these operating systems do offer features that make building and maintaining a website more productive. For example, there are numerous ways to access the administrative functions: the "old" Control Panel; an Explorer-like Computer Management, and a wizard-like Configure Your Server, which has ample help and support information. Another nice feature is the Microsoft Management Console (MMC), which makes it easy to access options.

Windows 2000 Server and Windows 2003 Server operating systems are faster, more reliable, heavier-duty, and easier to use than the older Windows NT operating system. Most of their more important server functions, such as, network, storage, and security have been centralized; although, TCP/IP and other Internet features are still not very well organized. Still with the introduction of Windows 2000, Microsoft at last made inroads into the Internet environment.

Note: The Windows 2003 Server Web Edition is available at a significant discount when compared to the Windows 2003 Standard Edition.

Windows XP Professional Edition

If you want to use a Windows product for your web server, your best bet is to use Windows 2003 Server Web Edition. However, if you must use Windows XP, get the Professional Edition (not the Home Edition) since like Windows 2003 Server the Professional Edition includes Microsoft's new Internet Information Server IIS 6.0 version, and support for dual processors. As long as you use the Windows XP Professional Edition along with secure open source and freeware tools (many are available via the Web), this operating system is adequate for a website's purposes. But note that Windows XP doesn't support Java, which may cause some problems. Also it may be difficult to use any web server other than the bundled IIS product.

Note: *According to a mi2g Ltd. report, for the twelve months ending September 2003 Linux was the most breached operating system followed by Microsoft Windows operating systems. The report states that 59.2% of all overt digital attacks were on systems running Linux and 20.8% were on systems running Windows. However, the report also acknowledged that the known, onerous difficulties with administering a Windows-based system — the effort that must be expended in 24/7 configuration management, reboot-patch-reboot (open source systems usually can be patched without the reboot drill, which is a valuable commodity in an e-commerce environment), etc. — when compared to open source solutions such as Linux and OpenBSD, the necessity of such hands-on administration could actually be serving to help Windows-based systems block many attacks.*

The report also found that the reason for most Linux breaches were due to poorly configured or updated third party applications and servers running on Linux systems. In fact, in mi2g anonymous interviews with hacker groups, it found that hackers prefer attacking Linux systems because of the plethora of configuration errors within those systems, which makes mounting attacks easier. However, mi2g also found that individuals hacking for intellectual gain or fun (rather than cybercrime) are generally more attracted to the perceived greater challenge associated with hacking well-configured BSD, Linux, Solaris and Unix systems (and any high-profile organization's online systems). Although the relative market share of different operating systems ought to be reflected in the proportion of digital attack suffered by each of them, the precise proportions fluctuate considerably — generally in a way that is coupled with the preferences of the hackers and the announced vulnerabilities still to be patched.

In the opinion of mi2g, there were several motivations and issues involved when a Windows-based system was targeted over Linux or vice-versa, "ranging from the reasons for an attack down to number of vulnerabilities being announced for Operating Systems, Servers, Applications and Libraries at a given moment in time and the ease with which the patches can be applied across a large scale deployment."

✪ YOUR WEB SERVER

Your web server is the software that serves your website's content to a customer's browser (Netscape, Internet Explorer, Opera, etc.). The web server software's basic functionality is quite simple — it takes a file name passed through a command, gets the file and sends it across the Internet so it can be viewed on the requesting computer's browser software. The web server software also tracks hits to the site, records and reports error messages, etc. Server-side technology is used to increase the functionality of a web server beyond its ability to deliver standard HTML pages — for example, CGI scripts, SSL security, and Active Server Pages (ASPs).

Today's web servers are used in such a variety of situations that it is necessary for the tools used to administer them to be quite sophisticated. Thus, more and more add-on options are available for web-related software development tools.

All of the major web servers, including Apache, Microsoft IIS, and Netscape Enterprise Server, have comparable functionality. This means that your choice of web server software may be based more on personal preference than on actual hard-core functionality considerations.

Server	September Count	September %	August Count	August %	Change
Apache	7,979,368	66.86%	7,705,536	66.94%	-0.08%
Microsoft IIS	2,853,576	23.91%	2,765,375	24.02%	-0.11%
Netscape	112,352	0.94%	109,898	0.95%	-0.01%
Zeus	101,128	0.85%	99,044	0.86%	-0.01%
WebSTAR	94,442	0.79%	91,315	0.79%	0.00%
WebSite	27,543	0.23%	28,979	0.25%	-0.02%
Other	765,715	6.42%	711,486	6.18%	+0.24%

Table 4. Results of the September 2003 Security Space Web Server Survey (11,934.124 websites responded), which is a survey of web server software usage on the Internet broken down by 86 domains. To find the latest Security Space Web Server Survey results go to www.serverwatch.com/stats/sspace/. *Courtesy of Jupitermedia Corporation.*

Apache

This is the most widely used web server and is probably the most stable and fastest of the web servers available. As Table 4 indicates, Apache holds the top position in the web server market space, and has almost three times as many installations as second place Microsoft IIS.

Apache was originally based on the httpd code that many say started the Web revolution. Prior to Apache 2.0 (the latest version), Apache was a largely Unix product that used a number of tricks in order to execute within other operating systems.

Apache 2.0, however, represents a major rewrite of previous Apache versions. Among the many changes, one of the most important is that Apache 2.0 supports a wide array of platforms in more efficient ways. Thus it is now possible to develop Unix- and Windows-specific execution models that make the best use of a specific operating system. This is because Apache 2.0 provides a new execution environment that separates the core functionality of the Apache system from the system that supports and processes requests. Thus Apache 2.0 is easily supported under a wide variety of operating systems: UNIX (of course), Windows (all versions), Linux, Mac OS X, BeOS, and more.

Apache web servers are released as "open source" with no fee for usage; there are also a lot of modifications and modules made for Apache. The source code for most operating systems (all versions of Windows, freebds, many versions of UNIX including Linux, etc.) is distributed in what is known as a "tarball." A tarball is just a source code directory tree that is packaged with the UNIX tar command and then compressed into a zip file. The most recent repository of Apache source code can be found at www.apache.org.

However, Apache isn't for everyone — setup and maintenance of the server requires familiarity with command-line scripting tools. Also, Apache lacks browser-based maintenance capabilities or GUI configuration/administration tools. Some users will be unhappy with the lack of visuals, Wizards, and browser-based administration tools.

Although there is no official support for Apache, the apache.org website is very useful. Bug reports and suggestions are distributed via the site's bug report page. Other questions can be directed to forums such as the one hosted by Serverwatch.com and Apache-server.com. Also check out newsgroups such as the one hosted by Google. Apache gurus can usually be found lurking about all of these sites. Be sure to bookmark not only the Apache.org site and Apache-server.com sites (they both provide a wealth of information), but also Apacheweek.com — this site is an essential resource for anyone running an Apache server. You can also find third-party companies that offer full commercial support for a fee.

Note: *If you would like a basic tutorial on how to install Apache, go to www.serverwatch.com/tutorials/article.php/1126341 for "The Newbie's Guide for Installing Apache." Bookmark the Serverwatch.com website — it provides a wealth of information on all kinds of server software.*

Microsoft Internet Information Server (IIS)

IIS receives high marks for superior installation, performance and maintenance. IIS especially earns kudos for its ease of installation and the quality of its management interface, which is provided separately from other interfaces. Another good feature is that IIS can be managed remotely via a web browser. However, IIS can operate only within a Windows environment.

The IIS 5.0 (included in Windows 2000) and the updated version, IIS 5.1 reflect only minor changes from the original IIS version 3.0 that came with Windows NT 4.0. But with IIS 6.0, which is bundled with Windows 2003 Server and XP Professional, there was an almost complete rewrite of this web server platform. IIS 6.0 sports a new execution model, better management facilities, and significantly increased performance. However, IIS 6.0 currently only supports Windows 2003 Server and XP Professional.

THE .NET FRAMEWORK

Some readers may have heard all of the buzz about the .NET Framework, but don't really understand what it is all about. Let's me see if I can give you a short explanation.

Microsoft .NET is software that connects information, people, systems, and devices. It spans clients, servers, and developer tools. The .NET Framework is Microsoft's new programming model for developing, deploying, and running XML web services and all types of applications-desktop, mobile, or web-based.

The .NET Framework is the infrastructure for the overall .NET Platform. The common language runtime and class libraries (including Windows Forms, ADO.NET, and ASP.NET) combine to provide services and solutions that can be easily integrated within and across a variety of systems. The .NET Framework provides a fully managed, protected, and feature-rich application execution environment, simplified development and deployment, and seamless integration with a wide variety of languages.

For a more detailed overview of .NET, check out Microsoft's "Getting Starting in .NET" web page, which can be found at www.microsoft.com/net.

Although this limits the deployment platforms for IIS-based web services, it also provides a number of benefits, including greater cooperation with the host operating system and easier management and control through a variety of standard OS tools and utilities. IIS 6.0, paired with Windows 2003 Server, offer admirable levels of integration with the .NET Framework.

Microsoft provides free online support through its knowledge base, which can be found at www.Microsoft.com. You can also obtain fee-based per incident service through the Microsoft Certified Support Center. If you purchase "support incidents" you can receive quick and knowledgeable technical support through an 800 number.

Netscape Enterprise Server (NES)

This web server runs a distant third in the web server market, but it still offers a "complete" web server package for any website. It provides support for the HTTP 1.1 protocol plus security enhancements in PKCS #11, FIPS-140 compliance, 128-Bit Step-Up Certificates, and Fortezza support. (Fortezza, Italian for "fortress," is a family of security products — PCMCIA cards, serial port devices, server boards, etc. - that are trademarked by the U.S. National Security Agency and used extensively by the military). It also comes with a built-in search engine, log analysis tools, advanced content publishing, server clustering and administrative rights delegation. It supports most Windows operating systems and most of the UNIX family including Digital UNIX, SGI IRIX, Sun Solaris, and IBM AIX.

NES is a good web server for a traffic intensive website since it has features such as SSL 3.0 support with client-side certificate authentication, SNMP and SMTP support and centralized server management. Technical support is free for the first 90 days only, then it gets a little pricey; you can opt for "per incident" based support or an annual subscription based support that ranges from $400 to $600. However, Netscape offers decent free support on its website (http://wp.netscape.com/enterprise/v3.6/).

Xitami

This is a freeware, robust, entry-level web server that will work with just about any operating system (although Mac support is not currently available). This web server is perfect for someone operating on a tight budget, especially if the hardware used for the web server is an older Pentium. Installation is simple. There are no wizards but there is a browser-based interface and command-line support that is efficient and intuitive. Documentation is thorough and makes the process of getting up and running understandable for both novices and pros. The website www.imatix.com/html/xitami/ provides extensive online documentation, help via email, and a link to a discussion group. Third-party support is available for a fee.

Zeus

This web server ranked fourth in the September 2003 Security Space Web Server Survey, and it may be just the ticket if you envision an e-commerce business with room to grow. Although a bit pricey (around $1400), Zeus is a scalable, secure, and high-performance web server. Zeus can be found underpinning business-critical solutions for web-hosting, content providers, and secure e-commerce companies.

What makes Zeus so popular is that it uses a small number of single-threaded I/O processes, which are capable of handling tens of thousands of simultaneous connections. Whereas the front-runner, Apache, uses a dedicated I/O process for each connection request, limiting it to 256 simultaneous connection requests.

But to handle the kind of I/O processing that Zeus offers you need either very fast equipment or a good cluster of servers. Fortunately, Zeus comes native with web server clustering support enabling a set of web servers to act as a single web server for the end user and allowing the load of serving web pages to be balanced across a set of different computers and (assuming your website has the bandwidth) multiple connections.

But unless you need the power Zeus offers, go with another web server because you will not realize any measurable benefit from Zeus over Apache (or another web server) unless your website experiences regular high traffic volume. Another possible downside to Zeus is that it runs primarily on Unix-based systems, and thus it is difficult to configure if you are not familiar with the command-line nature of the Unix family.

You can find extensive documentation and online support for Zeus products at http://support.zeus.com/.

Log Files

Okay, you've installed your web server software and even remembered to install and configure a firewall. It's now time for you to understand your web server log files.

Your web server creates records — called log files — of everything that happens on the server. Log files are actually huge files that contain a record of each and every activity that occurred on your web server. When I say "everything" I mean "everything" — every request for an HTML page, every graphic file requested, every request to have an active page executed, and every CGI script that ran. The web server considers each of these events a "hit." When the e-commerce industry first started ballyhooing the amount of traffic a certain site received, it was referring to the number of hits as tracked by the web server log files. Therefore, the definition of a hit, as far as web server log files are concerned is: Anything the web server is asked to do when servicing the demands created by the traffic flowing through its content, i.e., your website. Please note that the number of "hits" is not an accurate indicator of the number of people that are actually visiting your site. For the all-important "people count" or "unique visitor" statistics you need "log analysis software" to interpret website activity data.

As long as your web server has its logging feature activated, all of the activity on your website is stored in records. These records keep the details of which pages were requested, when they were requested and even information, such as, who initiated the request, the initiator's IP address and the type of browser used, along with how the initiator found your website. All of this information is stored in files named "Access Log," "Error Log" (such as a page that no longer exists) and "Referrer Log." By providing instant, ongoing, fairly exact and specific snapshot of the website's traffic patterns, these log files provide website owners with a plethora of raw data concerning their customers.

Because log files provide a record of all user activity, they have great value. You will even find the historical information useful. Maintain these files either in an encrypted form on the web server or store them on a separate machine offline.

Log files are important to the effective management of your website in three different ways. First, there's the overall "load" placed on the web server. For example, at any time did the site's activity exceed the capacity of the hardware and/or software? If so, this is a "heads up" that you need to improve your site's ability to handle the traffic peaks. Perhaps you need to upgrade the server hardware or install a higher-capacity web server software package.

Second, there's the measurement of the traffic or number of visitors that come to your website, what they look at, and how long they stay. This gives you good insight into what is succeeding or not succeeding on your web pages and in your marketing campaigns.

Third is security. The first line of defense against hackers is your log files. How? By monitoring your log files on a regular basis, you can spot suspect goings-on. You should also install trap macros (macros are small simple programs written to automate specific tasks) to watch for attacks on the server. While you are at it, also create macros that run every hour or so to check the integrity of "passwd" and other critical files. (The macros should be programmed to send email to the system manager if a change is detected.)

Log files expand very fast. If you are using an Internet hosting company to host your site, it will schedule these files to be rotated in such a way that older log files are regularly deleted. Believe me when I say you want to preserve these files. The easiest way is to institute a system and schedule wherein the log files are emailed to you automatically for organization and storage. This will guarantee that you will have immediate access to your data. However, this "raw" data is merely the tip of the iceberg and to be of any real use it needs to be analyzed — that's where log analysis software steps in.

✪ LOG ANALYSIS SOFTWARE

Web server log files are a potential treasure trove of information — they provide websites with a profusion of raw data concerning visitors to the site. Historically, log files were used most for a quick overview of bandwidth problems and basic tracking information. But although these logs contain an instant, ongoing, fairly exact and specific snapshot of your website's traffic patterns, they are somewhat difficult to read and to understand. This is where log analysis software is of help. Such software can take this raw data and tell you if pages are frequently or rarely visited. It analyzes how the data flows back and forth between your web server and your users, and gives information about IP traffic, e-commerce, cookies and browsers (vital if you're using Java or Active X). Log analysis software crunches the data to show how well your website is working, the origin and number of visitors, where they are going and how they found you in the first place. In addition, a detailed click stream analysis tracks every single click a user makes on your website. Such data is essential for auditing a specific user's activity. The more robust (and expensive) software can even offer historical and e-commerce reports reflecting time online and an authenticated user history.

The proper use of log analysis software data gives you valuable insight into how your website is being used. With some of the log analysis software available today, the easy-to-get-at statistics can be translated into slick graphs and bar charts for the non-technical crowd. However, unless taught otherwise, most website operators merely concentrate on figures such as the total of visitors to the site as a whole or to a particular page, and the number of banner click-throughs.

But log analysis software is also useful for finding problems with your site. For example you can track "error messages" that cause people to leave a site, if you see many of them in a day's report, you know you have a problem. You might see an error message and realize that the message only occurs with a certain browser version. Another example is security. Chapter 7, which covers website security, offers some useful ways you can use log analysis software to monitor your site and to help find culprits after a nefarious attack.

Large web-based businesses have astute marketing and advertising departments that use the results for trend analysis and use the extensive user demographics that such logs contain (i.e. how are users using the site). If you add to that user-registration and a responsible use of cookies, then the collected data can be transformed into a powerful customer intelligence tool.

Traffic Reporting

The traffic report segment of your log analysis software provides data on how much user traffic your website attracts, who's visiting the site, where the visitors are coming from, and how your web traffic changes on a hourly, daily, weekly, or monthly basis. A website traffic report is the simplest way to find out what works and what doesn't work. It can provide you with the information necessary to determine which web pages are not selected or to show a page that is immensely popular. Such reports could be compiled from the web server logs but it wouldn't be easy. In fact, without log analysis software it would be almost impossible to decipher details such as:

* The number of unique visitors to your site, when the came, from which countries, and from which referring sites (e.g. which search engine did they use to find your site)?
* What keywords did they use to find your website?
* Which pages did they access?
* How long did the visitor stay?

Understanding which outside sources generate page views can give you the opportunity to execute increasingly effective marketing strategies.

Usually your home page will be the #1 visited page on your site. By analyzing the traffic you can ascertain how your site design is succeeding. If you have a brochureware site, you'll now be able to tell if your potential customers go to the target pages. If you run an advertising site, you can compare the traffic on the web pages with a lot of banners to a page with fewer banners. If you operate a subscription site, you'll see what content is the most examined. If you have an online store, you'll know which products are the most scrutinized by browsing customers. If the sales figures don't reflect the same pattern, you now have a starting point to figure out why.

Conversely, if you find, via the traffic logs, that certain pages are less popular, analyze why. Traffic to an individual page may be influenced by your site design. Therefore, if it is the "About Us" page or "Investor Relations" page, don't sweat it. But if it's an important page that offers a new product or service, rethink the design. If traffic numbers to a specific area of your site disappoint you, look for possible reasons, such as technical problems. Perhaps you should provide a link on the home page or on a related content page. If you have an advertising site and you have a page that stands out as one that your visitors often use to exit your site, figure out why and fix it. Maybe there's a banner on the page that that causes a slow load time or an ad that is particularly irritating. The logs may show that you have traffic on the page but visitors come, see, and leave. They don't stay around and explore. Why?

After analyzing the traffic logs, doing your homework, and adjusting your website, wait a month or so and see if the changes helped. If not, try something else. It could be that your site is just fine but you need more public relations, advertising, or some other type of promotion such as a give-away. Again, after each little tweak, wait a month and analyze the logs. You'll gradually find what works and what doesn't.

Browsers and Operating Systems: Another aspect of log analysis software is that it can give you statistics on your customers' browsers and operating systems. (This data is usually found in your server's referer log files.) Take the lead from this data to ascertain if your customers use browsers sufficiently advanced so that you can add more bells and whistles to your site. If the majority of your customers are using the latest technology, you know you will lose little in traffic if you implement a Java application or a Flash plug-in.

How do Your Customers Find Your Site? This is important. You must know where your customers are coming from. Most should be referred to your website from search engines and directories. The second largest referrer should be from linked sites. Look at this data very carefully, if your top referrers are not Google or Yahoo!, find out why not. Is it something you can address? If so, do it. (Read Chapter 15, which discusses how to list your website with search engines and directories.)

Some log analysis software provide advanced reporting solutions such as monitoring web, ad, and streaming media servers, testing and checking broken links, and gauging page download times. This software also has the ability to monitor, to alert, and to recover server and network devices.

Note: So-called "packet sniffing" technology has been introduced into the traffic analysis market which eliminates the need to collect and centralize log file data. This technology gets its information directly from the TCP/IP packets that are sent to and from the web server.

The products available to help you analyze your website's usage vary as widely as the information various businesses request. The "basics" might provide you with the information you need without the sticker shock, if you are willing to work with limited options and don't mind that the presentation of data is "plain vanilla." More advanced products offer a variety of in-depth features and give you many customization options, including great reports with color graphs and bars.

The cost of the Log Analysis Software begins at "free" for AXS 2.0 (www.xav.com/scripts/axs/). This product can help you analyze the visits to your web pages. AXS 2.0 determines where visitors are coming from, charts their flow through your site, and informs you as to which links they follow when leaving. Also, you can analyze other information such as the referrer (Google, Yahoo!, MSN, Alta Vista, etc.), IP addresses, types of web browser, and the dates and the times of each visit. AXS 2.0 then processes the records into meaningful graphs and database listings.

Two low-cost options:

3DStats (www.3dstats.com) works with all web servers that use the common log file format. With 3Dstats, you can create a 3D bar chart showing the daily number of total HTTP requests (hits), the number of files sent (documents actually transmitted on behalf of requests), the number of "304's" ("Not Modified" responses caused by various caching mechanisms), and a number of other responses (redirections, not found responses, server errors).

Absolute Log Analyzer (www.bitstrike.com/analyzer) is a client-based log processing solution designed for web traffic analysis. This log analyzer product is easy to use and offers feature-rich tools that can generate fast, effective reports of any kind and display them in a format that's easy to read and understand. There are three versions of Absolute Log Analyzer - Lite ($50), Standard ($150) and Professional ($250). All ver-

Top Paths Through Site	Help 💡

This report tracks visitor activity beginning with their entry page into the site—the first page they open—then all subsequent pages during their visit. The default definition for a page in this context is defined as a document ending with the extension .htm, .html, or .asp. This definition can be changed by the system administrator.

Top Paths Through Site

Starting Page	Paths from Start	Visits	%

Homepage	1. Welcome Information
	    /
	2. Store Information
	    /store/
	3. Wireless phones View
	    /store/wireless_phones.asp
	4. W1000 Information
	    /store/wireless_phones/view.asp
	5. Add Product to Cart
	    /store/add.asp
		41	1.80%

	1. Welcome Information
	    /
	2. Store Information
	    /store/
	3. Wireless phones View
	    /store/wireless_phones.asp
	4. W1000 Information
	    /store/wireless_phones/view.asp
		30	1.32%

	1. Welcome Information
	    /
	2. Store Information
	    /store/
	3. Store Register Page
	    /store/register.asp
	4. Verify Information
	    /register/verify.asp
	5. Thank You for Registering
	    /register/thanks.asp
		24	1.05%

Figure 15. The WebTrends Log Analyzer's Top Paths Through Site report depicted in this graphic delivers a click-by-click view of your website's visitors' behavior — from their point of entry into your site, through the subsequent pages they visit, to the exit. *Graphic courtesy of WebTrends.*

sions have compatible workspace and database format, although the Lite version contains less statistical capability and has less flexibility while the Professional incorporates all available log analysis options.

Here are a couple of more costly traffic analysis tools you might want to consider:

Webtrends Log Analyzer (www.netiq/webtrends), which costs around $500, provides an easy-to-use website analysis solution specifically designed for the small business.

Sane Solutions's NetTracker Professional (www.sane.com/products/NetTracker/), which costs approximately $500, enables you to view the individual clickstream path of each visitor to your website, to analyze use of dynamically generated web content, and to measure site effectiveness of banner ad campaigns and response times and page delivery. The NetTracker Professional also measures site "stickiness" via repeat, new, and unique visitor behavior analysis.

Another option is a real-time monitoring and statistical reporting system such as WatchWise (www.watchwise.com). Each access to a registered website is automatically stored in real-time on the WatchWise server. It also monitors files that are downloaded from a registered website. Statistical reports are then generated in real-time, so you can determine who has downloaded your files. You can view these reports whenever you wish.

✪ DATABASES

A database is software that enables you to store data, to retrieve it, and to change it when necessary. With the appropriate database this is done easily and efficiently regardless of what the data consists of, or the amount of data you manipulate.

A database is integral to the design, development, and services offered by most e-commerce sites. To allow your customers to search your site for a specific product, you need a database. A database is necessary to collect information about your customers. A mailing list is fed from a database. Another example is that of web publishers who post up-to-the-minute information that is retrieved from a database. Get the picture?

Don't think you can build an *efficient, productive* website with its predestined multitude of web pages, without a database. If you try, your site will quickly become unwieldy. Templates, file systems and cut-and-paste can take you only so far. The simplest tasks, such as updating your product catalog, adding editorial content, and even maintaining simple links will eventually overwhelm you.

But take care not to be careless when implementing a database; first draft a strategy for feeding information into your web pages. For instance, look at the informational content of your web pages. Then decide how to implement more efficient ways of managing that content. How is your data currently being stored? What tools are available that can move it to the Web? Also be sure that the database interface is designed so that your employees easily can update the information with only cursory training.

A database can help you create web pages that display every single item in your inventory *and* keep each and every page current. The easiest way to do this is to create your web pages on-the-fly (or "on demand") with a program that "queries" a database for inventory items and produces an HTML page based on the results of that query.

That's just a tip of the iceberg. A database does more than simply provide users with access to information. A database can manage the website, keep links intact, and enhance the security of the site. Furthermore, it's actually much easier to maintain a database (once it's up and running) than it is to maintain a static website, with its many individual pages.

Build your e-commerce site around a database. You won't regret the decision. Now let's take a closer look at database technology.

Database Management Systems

Technically, a database management system (DBMS) is a software program designed to store and to access information used to support the workings of a website. A database can gather, handle and process information in an organizational structure to facilitate storage and retrieval of information. Once information is entered into a database, a DBMS can manipulate the information so that you can analyze it easily. There are many different types of database structures, but the majority of them organize information in the form of "records" and "columns," or "entries" and "fields." On a basic level, a database is somewhat like a set of spreadsheets with rows and columns.

A database will give you the ability to separate your content (your catalog offerings) from your HTML web page (the graphical design). It typically stores your catalog items in separate fields and tables as plain text with no formatting. An HTML template is then designed to provide a structure for the data as it is called from the database so it will be delivered to the website in a consistent layout every time. Databases allow your customers to quickly and to accurately search for what they want since the search is limited to named fields rather than a more expansive, full-text search.

Most brick-and-mortar businesses use databases, so, your IT department should already have experienced database managers. Still, database integration will present special challenges as you join your database with your website. That's because a website demands that the information fed into its page forms be extracted from the database (or moved from the user into the database) in ways that aren't necessarily a good fit for most brick-and-mortar rela-

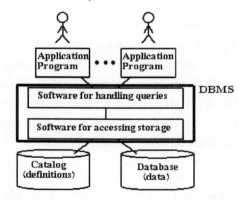

Figure 16. A database = collection of data; a catalog = definitions for database; database management system (DBMS) = software.

tional databases. Also, the rate at which a website's database receives hits usually far exceeds the norm found in traditional business applications.

Types of Databases

Before we get into the nitty-gritty of how a database models the data, let's look at the major differences between databases, differentiating between analytical databases and operational databases — since you'll often hear these terms bandied about whenever databases are discussed.

Analytical Databases: An analytical database, which is also referred to as "On Line Analytical Processing" (OLAP), is a static, read-only database, which stores archived, historical data used for analysis. On the Web, you will often find OLAPs used for inventory catalogs that hold descriptive information about all available products in a business's inventory. Web pages are generated dynamically by querying the list of available products in the inventory. The end product is a dynamically generated page displaying the requested information pulled from data residing in the database.

Operational Databases: An operational database, which is an integral part of what's called "On Line Transaction Processing" (OLTP), manages more dynamic data. An OLTP database allows you to do more than simply view archived data — you also can modify that data (add, change, and delete data). Typically, you will find OLTPs used when it is essential to track real-time information, such as the current quantity and availability of an item. As a customer places an order for a product from an online web store, an OLTP keeps track of how many items have been sold and when to reorder the items.

Database Models

A good way to conceptualize a database model is both as the "container" for the data as well as the "methodology" for storing and retrieving the data from the container.

There are basically three database models used today: text file, relational, and object. From the standpoint of simple information retrieval, virtually any kind of information can be plugged into a database (text, images, sound files, movies), but not all databases fit every business situation.

Text File Database

Although it's debatable whether a text file database is a true database, it is definitely not "relational." The best attribute of a text file database is its simplicity. Because of their simplicity and ease of use for the novice, you might be tempted to use this type of database for your website, but don't do it. Text file databases are not scalable and there is generally no concurrent access ability, i.e., two people can't use the database at the same time. For those with pre-existing text file databases, the most economical approach is to bite the bullet now and move the data to a relational or object-oriented database.

Relational Database

A relational database is a collection of "data items" that are organized as a set of linked tables from which data can be accessed or reassembled in many different ways without need to reorganize the database tables (a table is referred to as a "relation"). This type of database looks like a set of interlocking spreadsheets, with each table in the database being one spreadsheet. Relational databases allow simultaneous updates by numerous individuals. However, there are crucial differences — for example, all the data in a database column must be of the same type and the database rows are not ordered. Some of the best-known relational database packages include Microsoft's SQL Server, Oracle, Sybase, Informix, and IBM's DB2.

When designing a relational database, great care is given to "normalize" the structure of the data. The normalization process is performed by applying a series of rules to eliminate redundancy and inconsistency. The columns in all of the tables must depend upon a single key column with values that don't repeat.

A Relational Database Management System (RDBMS) is a program that enables you to create, to update, and to administer a relational database. Generally you are referring to a RDBMS when you talking about relational databases. An RDBMS typically takes statements written in the extremely popular Structured Query Language (SQL) and creates, updates, or provides access to the relational database. SQL is a "declarative" query language, which means that the user specifies what he or she wants and then the RDBMS query planner figures out how to get it. The RDBMS stores the data in whatever manner it "wishes."

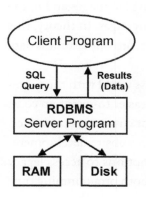

Figure 17. A typical client/server scenario consists of a user making a request of a database from his/her PC. Client software running on the user's PC establishes a connection across the LAN or WAN via TCP/IP sockets to another program (the database server) running on a more powerful, centralized PC. Once the connection is verified, the client software sends the SQL queries to the server, which plows through the data and returns the matching data back to the client software, which then displays the records on the user's PC screen.

Figure 18. For a RDBMS-backed website, the RDBMS client is the web server program (such as Apache) or perhaps a CGI script spawned in response to a user request for a URL. The user types a request or some other text into a form on a web client (Netscape Navigator or Internet Explorer) and that gets sent to the web server, which is itself an RDBMS client, or else spawns an RDBMS client (such as a Perl script) which has or opens a connection to an RDBMS server (such as Oracle). The retrieved data then travels back from the RDBMS server to the RDBMS client, to the web server, which relays it back to the user's web client.

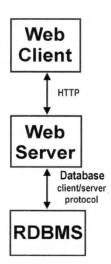

You need an RDBMS if you have data that changes frequently. Relational databases have *concurrency control* to ensure that the tables won't get corrupted even if many people simultaneously write to and read from the database.

The term "client/server" was devised to describe how users work with relational databases.

If you choose to back your website with an RDBMS, make sure your web server software can connect easily and deal efficiently with an RDBMS. The web server software mentioned in this chapter are all multithreaded, Tcl-enabled dynamic web servers designed to handle large scale, dynamic websites, i.e. they can deal efficiently with any RDBMS.

Object-Oriented Database

Generally, when talking about object-oriented databases (ODBMS), most people are in reality talking about an object-oriented database management system. ODBMS is a data management product that is specifically designed for use with an object programming language and/or is closely coupled with one or more object programming languages (C++ or Smalltalk). An object database, therefore, provides database management system functionality to object-oriented programming languages and to the objects that have been created using those programming languages. In theory, this approach unifies both application and database development into one seamless data model and language environment.

Object-oriented databases have long tantalized the business world with the promise of less application code, more natural data modeling, and easier-to-maintain code, but they have taken considerable time to catch on in terms of customer acceptance. While ODBMSs slowly have begun to move into mainstream website development, they still lack scalability and there is limited availability of options to access legacy relational databases. Object-oriented databases may improve both programming time and response time, but you must be willing to go for a whole package — to write in an object-oriented language, to switch to a new database, and to accept the inherent risk of doing things in a new way. But due to two features unique to object-oriented databases — objects can be heterogeneous (each can contain a different collection of "owned" data) and objects can contain some inherent "intelligence" — some e-commerce sites may find that it pays to go the ODBMS route.

Object oriented databases include: GemStone, NeoAccess, ObjectStore, and Fast Objects, to name just a few.

Hypothetical Relational Database Model

PubID	Publisher	PubAddress
03-4472822	Random House	123 4th Street, New York
04-7733903	Wiley and Sons	45 Lincoln Blvd, Chicago
03-4859223	O'Reilly Press	77 Boston Ave, Cambridge
03-3920886	City Lights Books	99 Market, San Francisco

AuthorID	AuthorName	AuthorBDay
345-28-2938	Haile Selassie	14-Aug-92
392-48-9965	Joe Blow	14-Mar-15
454-22-4012	Sally Hemmings	12-Sept-70
663-59-1254	Hannah Arendt	12-Mar-06

ISBN	AuthorID	PubID	Date	Title
1-34532-482-1	345-28-2938	03-4472822	1990	Cold Fusion for Dummies
1-38482-995-1	392-48-9965	04-7733903	1985	Macrame and Straw Tying
2-35921-499-4	454-22-4012	03-4859223	1952	Fluid Dynamics of Aquaducts
1-38278-293-4	663-59-1254	03-3920886	1967	Beads, Baskets & Revolution

Hypothetical Object Database Model

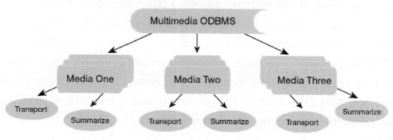

Figure 19. A relational database consists of a set of tables, where each table consists of a fixed collection of fields. An indefinite number of records occurs within each table. However, each record must have a unique primary key, which is a sort of name for that particular bundle of data. Objects in an object-oriented database are bundles of data and behaviors. Because objects in an object-oriented database can contain a variety of attributes and data, queries are often performed through a set of methods. Each object implements these methods in a way that is appropriate for itself. In the example given here, two methods might be "summarize" and "transport." *Graphic courtesy of IBM.*

Differences between Relational and Object-Oriented Databases

Relational databases monopolized the business world until the introduction of object-oriented databases. The primary differentiation between relational and object databases is that relational databases are built on a row/column paradigm, similar to a spreadsheet, while object databases treat every item on the website — from graphics, to audio files, to URLs — as separate objects. This differentiation means that rows of flat text map well into a relational database structure, but images and sounds do better in an object-oriented environment.

Relational databases are terrific for storing data that can be converted easily into a two-dimensional representation (i.e. text). However, not all objects found on the Web are amenable to such structuring (i.e. images and sound). Indeed, "flattening out" complex data structures to fit into tables sometimes resembles fitting square pegs into round holes. Thus to use a relational database backend for storing images and sounds, you need special coding that allows the object to be mapped into the entries of the relational database. This extra coding not only puts an added strain on your site's processing power (speed), it also has the potential to introduce intermittent problems that might have otherwise been avoided.

An RDBMS also uses the intersection of rows and columns for every instance of an entity in the database, regardless of whether a particular cell is needed in a particular instance, while an ODBMS only stores the particular parts of the data actually used in any specific instance.

Unlike a relational DBMS, object DBMSs have no performance overhead to store or retrieve a web of interrelated objects, such as a set of interlocking tables. A natural one-to-one mapping of object programming language objects to database objects allows for higher performance management of objects as well as better management of the complex interrelationships between objects. This makes ODBMSs better suited than RDBMSs to support website document structures, which have complex relationships between data. When everything is working correctly, object databases can operate on complex data, such as images, multimedia, and audio, as efficiently as a relational database operates on simple text or data.

Final notes:

*   If you offer images or sounds as a main feature of your website, consider using an object-oriented database.
*   Relational databases generally store all of the data on a single server, while object databases can call upon multiple servers.
*   It is possible to build a website that uses both relational and object databases in a

hybrid system. The relational database can store data such as text and software downloads and the object database can be used for sound, video, or images offered up by the website.

Desktop Database Software

While not a viable solution for most e-commerce businesses, if you are willing to tolerate limited capabilities and performance and are operating on a shoestring budget, you can build a small website using a desktop database software offering. There are three desktop relational database solutions that have some built-in web capabilities, although none come near the quality and performance of a high-end solution. Still, these three products — Microsoft's Access, Corel's Paradox, and will, with varying degrees of success, let you get your data up on the Web quickly and inexpensively — as long as you allow for some tweaking here and there.

* Microsoft's Access allows you to create a database with its Wizards. It is adequate when it's enveloped by and communicates only with other Microsoft products; but it isn't happy when forced to interact with other manufacturers' software.

* Corel's Paradox allows you to create a database with its "Experts" (which are similar to Microsoft's Wizards) and like Microsoft's Access, Paradox too requires much effort to interact with other manufacturers' software.

* FileMaker Pro is the best of these three products. It is easy to use and has good documentation and templates that will allow you to build basic databases. It is one of the simplest and least expensive relational databases you can buy.

Overview of Robust Database Management Systems

Vendors offer a variety of database management systems that are suitable for e-commerce. Each system has its own strengths, capabilities, and challenges. Here are three of the more popular system:

MS-SQL is a DBMS that is fast, scalable, and free. It offers ease of administration along with numerous automated features (available only for the Windows NT platform) — but there is a trade-off — much of the SQL Server is not user definable with almost everything decided automatically by the system. In cases where the system makes the right choice, this feature works transparently and saves time. However, in cases where the system makes the wrong decision or choice, it can be a nightmare. Following along the automatic configuration idea, Microsoft has tried to create a self-teaching database that examines and adjusts itself to satisfy whatever requirements are being placed on it at a given time.

ObjectStore Enterprise Edition is a reliable OBDMS for high-performance applications built for delivery of high-speed, complex web transactions, and dynamic content needed by today's web-based businesses. ObjectStore can shorten development time dramatically by reducing the amount of complex code required.

The Oracle DBMS provides a wealth of application possibilities along with increased manageability and improved security. With the JServer option Oracle provides a runtime environment for Java objects and with the WebDB option, Oracle provide the ability to manage the content-creation process while also distributing creation tasks to reduce resource bottlenecks. Oracle addresses the security issue through its "virtual private database" support that allows the attachment of security policies directly

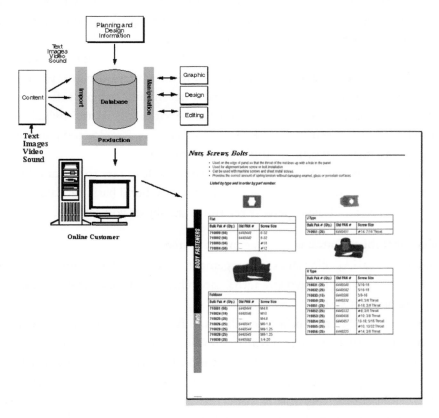

Figure 20. All of your website's content is stored in a database. To "publish" that content on your website, value is added through the use of tools for manipulation, refinement, and presentation. The production of the content is then achieved by selecting, extracting, formatting, and postprocessing the content for specific web pages.

to tables and views, thus the same security policy is in place regardless of which application is accessing the data.

✪ DATABASE PUBLISHING

While on the subject of databases, let's discuss how you might use a database to create and to maintain your website's content. While you might be feeling a little overwhelmed by now, setting up your website with a database is a big step towards building a flexible, scalable site. Stay with me as I try to simplify things a bit. There are two types of web pages — static and dynamic. They both have the same goal — the use of information stored in databases.

The action of a website that collects data from a database and then builds pages from that data is referred to as "database publishing." With the right interface, even a novice user can go into a database to update information that can be made available immediately.

Maintenance is easy in the database-publishing world — for example, one can create an HTML template once and merge that template with new content to provide a reliable way to put information on your website in a consistent layout.

Security is also easier since databases allow you to keep the wrong content off a website through the enforcement of security checks, such as "hide" flags, checking the date against the date a document is set to be posted, the date a product is set for delivery, etc.

Though database publishing can help you manage your website, it may take some time for you to learn all the ins and outs of the related software tools. Database publishing requires not only a database, but also it may require extra web server software and an application server.

Note: When using database publishing software be sure to perform a number of tests before going "live." During the tests, check for slow pages, heavy pages, small pages, missing graphics, etc.

Static Database Publishing

The static website's information is already prepared in the desired format for use by the customer. Some website operators will begin their venture into e-commerce by creating a simple static website paired with a database. This technique is often referred to as "static database publishing." With static database publishing, a visitor can't search through the database on demand, nor perform many of the fancier dynamic-site tricks. But, by using a minimal database approach, you ease the burden of data management, while leaving open the opportunity to move to a full-blown dynamic database-driven site later when business demands a more dynamic environment.

Note: A static website displays infrequently changing information or data. It's like putting a snapshot of your business' information onto your website. The contents of a static site can be marketing brochures, white papers, monthly newsletters, software, or even a small product catalog, etc. This information is not interactive, nor does it change very often.

Static database publishing "automatically" generates static HTML pages from a database, although changes to the database will only show up on the website when new pages are generated manually and uploaded to the web server. The advantage of this type of website content management is that you need very little technical expertise and the site is easy to maintain because static database publishing can be performed with any web server — there is no requirement for special features such as CGI scripts. Thus, if you are using a web-hosting service for your website, you will always be able to use this method irrespective of who your web-hosting provider may be.

The disadvantage of static database publishing is 1) the manual interaction required to upload the newly generated files to the web server and 2) the ability for visitors to update their information isn't available.

Tools for Static Database Publishing:

DBtoHTML (www.xlinesoft.com). This product can be used to generate web pages automatically from any ODBC compliant database making it an excellent tool for creating static HTML pages from a database. Since DBtoHTML doesn't allow you to create an overview of the database's full contents (and it has no search function), this product is better suited for databases with a few large records. Cost — $129.

Note: Open Database Connectivity (ODBC), which is a Microsoft standard database access method, is a database common interface. ODBC, in simple terms, inserts a middle layer (a database driver) between an application and the RDBMS to translate the application's data queries into commands that the RDBMS understands. For an ODBC interface to work, however, both the application and the RDBMS must be ODBC-compliant.

GDIDB Professional (www.gdidb.com). This product allows you to publish any ODBC compliant database (or spreadsheet) to your website. GDIDB Professional enables you to create a large easy-to-maintain website by separating the site design from the site content since you keep the content in a database. You can even design your website so its content is driven from your live business database. With GDIDB Professional you can format your content on your web pages any way you want it, create any HTML link structure that you can imagine from your relational database, publish web pages at the click of a button, minimize upload times by only uploading files that have changed since the last publication, and automatically publish your database to a pre-set schedule of unattended publishing. One feature that is particularly interesting is the ability to add form data automatically to the database (with some additional help from you). Cost — $76 to $220.

The best scenario in which to use static database publishing is when you have a website that has static content, i.e. it is not changed often. Many brochureware sites fit this scenario.

Dynamic Database Publishing

Dynamic database publishing differs from its static counterpart in that dynamic publishing requires the use of certain parameters or keys to search a database in real-time. Then when the desired data is found, a web page is created for the customer's use. Dynamic database publishing also requires Java scripts, CGI programming, ISAPI, NSAPI, and more. ColdFusion MX, Lasso, and WBSP are some of the more popular application server software used to transform a database's data into dynamic web pages.

A dynamic website can be an online store with an ever-changing product catalog

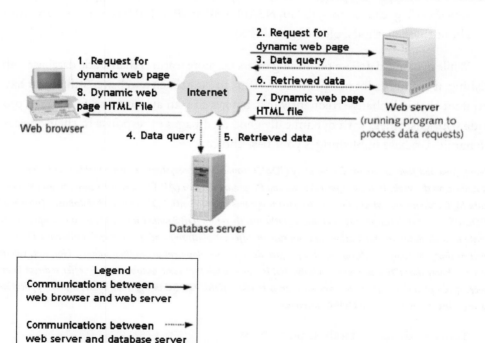

Figure 21. This graphic shows a typical interaction between a customer's PC and a website when the website is built upon a database-driven architecture.

so that when a customer requests information, a database can provide not only the product but also the current price, colors, sizes and the availability. Car dealerships, real estate, and ezine (online magazines) sites use dynamic data.

Dynamic database publishing requires that the website be built upon a database-driven architecture so the website can produce web pages that instantly respond to a customer's query. That means such websites are more difficult to implement and use than those built via static database publishing techniques because they require that you and/or your staff possess more technical knowledge and capabilities. There are also other downsides to building a web-based business around dynamic database publishing. They include:

* Unless you carefully design and institute a good maintenance system, a large volume of traffic on your website can cause data traffic congestion and performance problems.

* The web server needs additional features to support dynamic database publishing. For instance, in order to create dynamic pages your web pages must be able to interact with a supporting program, script, or applet and need to be supported by the web server via, e.g. CGI scripts, ISAPI, NSAPI, ASP, PHP or JSP (For more information about these methods, please see Table 5).

While dynamic database publishing offers far more features than static database publishing, it requires a lot more time and effort to build such a website. Still, if you have content that will change frequently (e.g. a news site, an auction site, or want to provide your customers with dynamically generated pages in response to their queries) dynamic database publishing is your best choice.

Note: Java and Java Database Connectivity (JDBC) from Sun Microsystems is also a viable way to publish a database on the Web. JDBC is a Java Application Program Interface (API) that enables Java programs to execute SQL statements, which, in turn, let Java programs interact with SQL-compliant databases (nearly all RDBMSs support SQL). Java runs on most platforms, therefore JDBC makes it possible for a web operator to write a single database application that can run on different platforms and interact with different RDBMSs. At this time, the tools and drivers needed for Java database development are in their infancy. However, you do need to know that JDBC is a lot like ODBC but is new enough that some databases don't offer a direct interface. If you plan to use Java now, one solution is to use a JDBC-ODBC bridge that allows a Java application using JDBC to connect to an ODBC database.

Tools for dynamic database publishing:

Baserunner (www.baserunner.com). This product creates data-driven, dynamic web applications from any information stored in a dbf format database such as FoxPro

CGI   **Common Gateway Interface**
A standard method of extending web server functionality. CGI works by executing programs or scripts on the web server when a web browser request is received. CGI lets you create web pages that can return information based upon a customer's input by calling a compiled C program or Perl script to access a database or another data source. However, CGI is inefficient, since it causes the server to launch a new process to run executable programs every time a new user makes a request. If you have a lot of traffic, your website's performance will suffer. Still, CGI is a tried and true interface and is the most common way to connect with web servers. Here's an example of how CGI might work: A web browser sends form data to a CGI script on the web server; the script integrates the data with a database, and sends back a results page as HTML.

ISAPI   **Internet Server Application Programming Interface**
A web server application development interface that can be used instead of CGI.

NSAPI   **Netscape Server Application Programming Interface**
A web server application development interface, developed by Netscape Communications Corporations.

ASP   **Active Server Page**
This Microsoft technology dynamically creates web pages with an .ASP extension using ActiveX scripting (usually VB Script or Jscript code). When a web browser requests an ASP page, the web server generates a page with HTML code and sends it back to the browser. Thus ASP pages are similar to CGI scripts, and are usually used when it is desirable to enable Visual Basic programmers to work with familiar tools.

JSP   **Java Server Page**
Sun's Java Server Page (JSP) technology controls the content or appearance of web pages through the use of servlets (small programs that are specified in the web page and run on the web server to modify the web page before it is sent to the requested user). JSP is comparable to Microsoft's Active Server Page (ASP) technology.

PHP   **Hypertext Preprocessor**
PHP is a script language and interpreter that is freely available and used primarily on Linux web servers. PHP (originally known as "Personal Home Page") is an alternative to Microsoft's Active Server Page (ASP) technology. As with ASP, the PHP script is embedded within your web pages along with the HTML. Before the page is sent to a requesting user, the web server calls PHP to interpret and perform the operations called for in the PHP script.

Table 5. Detailed explanation of how a website using dynamic database publishing might communicate with the database to generate the necessary dynamic web pages.

THE ROLE OF APPLICATION SERVERS
IN DATABASE PUBLISHING

Some e-commerce website operators will opt for a different method to enable dynamic database publishing; they will use application server softer to transform the database data into dynamic web pages. To achieve this you must install software such as ColdFusion MX (www.macromedia.com), Lasso (www.blueworld.com), or WBSP (www.whizbase.com) to pull information out of the database to create HTML pages on demand. This, in turn, means that you must run not only a web server and a database server, but also an applications server.

and dBaseIV (a popular non-relational database program). As such, baserunner provides a fast, inexpensive, small-scale solution that permits simple template-based development and avoids the overhead associated with such technologies as ODBC and the complexity of SQL, PHP, or ASP. (It is noted, however, that Baserunner 4.6 introduced support for relational databases and for using multiple tables within a template.) The best thing about this template-based, stand-alone CGI executable is that it provides flexible control of your online data. It works as an extension to HTML, providing new tags and processing directives that enhance your web design by giving you easy access to dynamic content. While this makes it a relatively easy-to-use tool, it also requires that you possess more than a bit of technical knowledge to use it to its best advantage. Cost — Free.

KazTrix DataBuilder 1.0 (www.instabase.com). This product allows you to publish data from about any program (including personal information managers) or database (including Microsoft's Access). It is easy to use, and although it gives you many of the advantages of dynamic database publishing, it only requires the user to possess minimal technical knowledge. KazTrixR DataBuilder requires no CGI scripting or any other plug-ins on the server side. However, it is also not customizable and it lacks some of the advanced features found in other dynamic database publishing products. Cost — $149.

Dynamic Websites and the Search Engine Dilemma

A search engine has three parts. The first is a spider (also called a "bot") that goes to every page on every website that wants to be searchable and reads it. Every search engine has its own criteria that its spiders use when crawling a website. The second component is a program that creates a huge database of the keywords in the pages that have been read by the spider. Finally, there's an access program sitting on a portal that can take millions of search requests from users, compare them to the entries in the database, and return results.

Although dynamically generated web pages are easier for the web operator to manage, they are difficult for some search engines to index and incorporate into their database. The simplest answer to this problem is to use static pages as often as possible. Use the database to update the pages instead of generating them on the fly. Another little bit of advice: Search engines hate the "?" symbol, so don't use it in your URL.

✪ CONCLUSION

Before launching an e-commerce website, take the time to understand your website's infrastructure. This includes the operating system, the web server, the log analysis program, and the database (an integral part of most e-commerce sites). And, of course, the hardware on which these applications run.

Overall, deciding which operating system and web server to use is just a matter of personal choice, but the log analysis software and database choices can make or break an e-commerce business. Log analysis software points out what works to your advantage along with any trouble spots. And the database provides a simple, flexible, reliable, and affordable solution for supporting a large volume of data. This data can include not only merchandise, but also such features as online membership roles with login, logout, and expiry date control. If you already have an operational e-commerce site and find that you are spending too much time managing its contents, a database can help.

In the end, most e-commerce operators will find that databases are what provide their customers the accessibility they demand. When product catalogs, online services, automated email responses, user feedback systems are managed via a database everyone is happier. Thus plan your website from the get-go so that a database can be an integral part of your operations — now and/or in the future.

Chapter 9

E-Commerce Solutions

*Conspicuous consumption of valuable goods is a means of reputability
to the gentleman of leisure.*

—Thorstein Veblen,
The Theory of the Leisure Class

While some e-commerce operators will choose to handle all the tasks needed to build and design their website, others will choose to outsource some or all of those duties. And, to be honest, if money is no object, you should hire a consultant or web-designer to help in the building and design stages. However, most readers' budgets can't accommodate the costs of outsourcing such chores and thus they must take care of the website design aspects themselves. This means that they are charged with the task of finding the best service or e-commerce software package to meet their budget and requirements. This chapter is written for that group, but even those with deep pockets should read this chapter to understand what their outsourcer(s) offers.

In the previous chapters we discussed the necessities of a basic website infrastructure, now its time to determine how to incorporate that infrastructure into a working e-commerce site. This requires some sort of e-commerce software. The application or program you chose must enable your website to provide your customers with the convenience of on-line shopping, i.e. provide your customers with the ability to view your goods, to add or to delete items from their selection, and to review their final selection prior to purchase.

Whether you are a novice or an expert, whether you are starting a new web-based business or giving your brick-and-mortar business a web presence, there is an e-commerce software package that is right for you. It doesn't matter if you have just a few products or a whole catalog that gives your online customer thousands of choices, determining the right e-commerce software package for your e-business can be a confusing undertaking. For example, some e-commerce products are designed to grow with your business, but other products work well only in low traffic environments. Use your blueprint — it will make the task easier.

One of the first decisions to make is whether to enable online order processing or handle your orders via an order form that is then emailed to you for processing. Next, you must decide whether you would rather run your website from an e-commerce package, an "out of the box" solution, or a "software toolkit," which requires additional programming to provide a complete, but flexible, customized solution. Here again, use your blueprint — it concisely lays out your site's exact needs and specifications.

Note: Consider your own and your staff's "expertise" with computers before making a final decision as to which e-commerce solution is right for your e-business.

While the author has categorized the various e-commerce packages, please note that these categories often spill over into each other. However, there are a few items that are consistent throughout the categories and as you examine the various e-commerce solutions, you should consider:

* The product's ease of set up.
* Will the software integrate easily with your existing software?
* Is scalability limited to what is offered by the hosting service or can you integrate new software products?
* How are orders processed?
* Is the software difficult to administer?
* Can it import product data from a database?
* How thorough and responsive is the documentation and support?
* Will your website operator have limited technical knowledge? If so, the software should make extensive use of wizards and templates.
* Or, if you have an experienced website operator, the software can be more flexible and responsive to your needs such as software that can create CGI scripting and HTML pages.

Note: It is of the utmost importance that you install your online e-commerce software on a Secure Sockets Layer (SSL) server to facilitate safe (and private) online transactions.

✪ HTML EDITORS

Know the difference between web design software and e-commerce software. Web design software (commonly referred to as "HTML editors") provides you with the basic tools for designing a website. Only those readers developing a brochureware site or going the email order processing route should consider using this type of product. For those readers taking one of these paths to e-commerce, any of the following products should suffice:

CoffeeCup HTML Editor (www.coffeecup.com). This software makes it easy for even a novice to create a website. It comes with more than 40 website templates, 125 JavaScripts, a DHTML menu builder, numerous graphics, wizards for frames, tables, forms, fonts, and more. It also provides helpful and easy-to-follow documentation on how to use the software. Cost — $70.

EZGenerator (www.ezgenerator.com). Working with this software is almost as easy as working with a word processor. Just type, select, drag-and-drop your pages to build your website. The product provides 2000 template variations so you can change the look and feel in one simple click of the mouse. It also automatically (well almost) creates your website navigation for you as you create new web pages and categories. Cost — $100.

HotDog Web Editor (www.sausagetools.com). This is one of the most popular HTML editors. The HotDog Pro 7.3 provides a comprehensive set of web editing tools including Linkbot, Interactor, and Paint Shop Pro. As such, the HotDog Pro provides the web designer with the complete tools for the creation and management of a professional website. But it still offers intuitive features and customizable tools so that even the novice user can create great looking websites. Cost — $100.

Microsoft's FrontPage 2002 (www.Microsoft.com). If you are a first time web designer, this may be the product for you. Frontpage 2002 can help the novice to get their website up and running easily and quickly, especially if the designer is familiar with the Microsoft Office suite of products since it uses some of same buttons and commands. The WYSIWYG (What You See Is What You Get) interface is invaluable since you don't really need to understand HTML codes and scripts. This turnkey product also makes its easy to add DHTML effects (changes on demand are also quite simple). The product has an insert form function, too, which makes adding forms for emails and guest books a breeze. The primary problem with Frontpage 2002 is that if you aren't hosting your website yourself, you may have to work a bit harder to find a web-hosting service that allows FrontPage extensions. Cost — $150.

SiteDesigner 2.0 (www.sitedesigner.com). SiteDesigner is a WYSIWYG website authoring tool, that has a drag and drop editor. This makes SiteDesigner extremely easy to use. Once you set up a master page for your site (using the templates that are included with SiteDesigner, or using your own images), every web page you create thereafter can inherit that master page's properties. The SiteDesigner generates extremely clean HTML code (some WYSIWYG editors add a lot of unnecessary code meaning pages are harder to troubleshoot and take longer to load). Cost — $100.

Stone's WebWriter 3 (www.webwriter.dk/english/index.htm). This award winning Danish no-nonsense HTML editor offers easy-to-use dialogues, right-click editing, code completion, and syntax high-lighting so you can design your website without knowing HTML. WebWriter is for the professional as well as the beginner. With Web-Writer you can build JavaScript, Cascading Style Sheets, and image maps. Cost — Free (private use) to $800 (for a large enterprise that needs unlimited licenses).

✪ SHRINK-WRAP E-COMMERCE SOFTWARE

Now let's discuss software that will provide you with all the tools you need to build a full-fledge e-commerce website, e.g., a product catalog, a shopping cart, and some kind of order processing and transaction security solution. E-commerce packages of this type enable you to get your website up and running quickly. However, be aware that when a product touts "ease of use," it is a subjective term — what one person considers an easy-to-use product, could be viewed as difficult by another. But if you know HTML and have CGI experience you may want to use a shrink-wrap product; they all provide the web design with some degree of flexibility, such as designing your own HTML pages.

Any shrink-wrap e-commerce solution should allow freedom of choice to incorporate software from a variety of offerings. Furthermore, they should provide, at a minimum:

* Good documentation and support.
* The ability to import data from a database file.
* Order processing features such as the availability of a virtual shopping cart.
* The ability to transfer data securely using SSL and not leave it in an unsecured area of a server where unauthorized parties might find it.
* The ability to send customers' details to you using encrypted email.
* Simplified day-to-day operation of your website such as allowing changes to be made offline and then to be uploaded to the server. (For security reasons, this feature usually requires that you only use one specific computer for the updates.)
* The ability to add, to delete, and to amend product data as well as to run special promotions.

✳ Good, detailed reports of the analysis of server logs, such as the number of hits and referrer information which in turn can give, for example, a sales history analysis and information about the most common entry and exit points your customers use in your website.

The majority of the shrink-wrapped e-commerce packages will have in addition to the features set forth above:

✳ The ability to accept orders and payments in as many ways as possible — credit cards, debit cards, paper checks, electronic checks, digital cash, fax, telephone, or snail mail.

✳ Maintain pre-set tax tables that so the correct tax is collected on each order.

✳ The ability to interface directly with carriers such as UPS, along with automatically calculating shipping costs.

✳ The ability to send an email order acknowledgement automatically to the customer along with a unique number to facilitate order tracking.

You will also find some of the products offer services such as:

✳ Domain name registration.

✳ Automatic search engine submissions.

And more advanced features such as:

✳ Autoresponders (mail utilities that automatically send a reply to an email message).

✳ Chat rooms.

✳ The ability to handle online processing easily. (Although you might process your orders offline, it is good to have this flexibility for future growth.)

✳ Discount clubs that let you give discounts to repeat or high-volume customers.

✳ Online order tracking that lets your customers check the status of their orders.

✳ Inventory management facilities, which can remove a product, automatically, when supply dips below a certain level.

✳ Additional marketing tools such as the maintenance of customer-buying history and preferences, targeted emailing capability, and affiliate program management.

Be aware, however, that many shrink-wrap e-commerce solutions lack support for the fundamentals, such as back orders, and some lack the modularity needed to mix and match applications from competing platforms. Now let's examine a sampling of just a few of the many products available.

AceFlex B2C (www.aceflex.com). Although pricey, this robust sales management software gives you everything you need to establish an e-commerce site. For example, the AceFlex B2C provides web-based sales tracking, order processing, customer relationship management, product/content merchandising, scheduling & analyzing promotion/marketing campaigns, set price levels, easy customizable storefront with advanced catalog, shopping cart, real-time shipping gateways, pre-integrated online payment processors, and more. Thus this product enables an e-commerce business visibility into its entire multi-channel sales strategy so your staff can discover and respond quickly to any customer's needs, thereby strengthening customer relationships. Also, it is possible to organize efficient collaboration with suppliers online. This translates into greater insight into efficient business practices, increased profitability, and better customer satisfaction. To check out the variety of websites that have been built with the AceFlex B2C solution visit: www.pdfstore.com (software downloads), www.adasa.com (apparel sales), www.karaokecdgmusic.com (offers streaming audio), and www.reach4life.com (mixture of content and sales). Cost — $2000.

Adobe GoLive (www.adobe.com). This product is actually CyberStudio 3.0 with a bit more window dressing, which isn't a bad thing. Of all of the products listed in this section, GoLive has the cleanest, most user-friendly interface. That's because it enables the designer to layout pages on either a grid (which is structurally dependent on complex automatically nested HTML tables) or via a simpler method of user-defined tables consisting of columns and rows. GoLive makes it easy to switch between the WYSIWYG layout mode and a syntax-checking source code view. Dynamic HTML (DHTML) and the related cascading style sheets (CSS) are easy to create and GoLive offers some unique timesaving design features such as automatically generating low resolution black-and-white images that load before a high resolution color image and it enables you to create JavaScript rollovers quickly. Cost — $400.

ecBuilder Pro 6.0 (www.ecbuilder.com). This award-winning website builder and e-commerce software program is suitable for any type or size of e-business. For instance, you can use ecBuilder Pro to construct a website and online store with real-time secure credit transactions, or an online store that can integrate seamlessly into an existing website. ecBuilder Pro 6.0 also enables you to build a brochureware website. Cost — $397.

Erol3 (www.erolonline.co.uk). This is a very flexible e-commerce software and shopping cart builder with excellent documentation. For examples of some of the websites built with the Erol product visit www.officewigwam.co.uk, www.kiddicare.com, www.chaseav.co.uk/erol.html or any of the numerous other websites that Erolonline features on its website. Cost — £99 and up.

Macromedia Dreamweaver MX 2004 (www.macromedia.com). This popular product is designed for the e-commerce business that needs the latest and greatest in website development. However, it's not for the novice — it has a steep learning curve and tech-support is expensive once the 90-day free period runs out. Nonetheless, this extremely powerful web development application offers many advanced development tools for website design and maintenance. For example, it provides support for cascading style sheets, provides a good interface for accessing HTML code directly, and offers features for the newest Active Server components. Cost — $400.

Note: A word about DHTML and cascading style sheets — both are handled differently by current versions of Netscape Navigator and Microsoft Internet Explorer and aren't supported at all by most older web browsers, whether Netscape, IE, or Opera. So when you use DHTML and cascading style sheets, you must use tags that are browser-specific and create scripts that redirect older browsers to alternative versions of your website.

The author's main gripe about many of the current crop of shrink-wrap e-commerce solutions (especially the less costly products) is that they can't integrate with database servers easily. Hopefully someday a budget-pleasing turnkey solution will be made available that will tap into databases easily and credibly convert existing database forms into web-ready HTML.

Since this book gives advice on what is needed to build a viable e-commerce business, but doesn't walk the reader through the actual website build, some readers will need more help when they begin the actual design or build process. This is where website design books, of which there are many, come into play. The author suggests: *Learning Web Design: A Beginner's Guide to HTML, Graphics and Beyond*, Jennifer Niederst (O'Reilly); *HTML: A Beginner's Guide*, Wendy Willard (McGraw-Hill); and *HTML: Your Visual Blueprint for Designing Effective Web Sites*, Ruth Maran (Wiley). In addition to these books, there are numerous websites that you might find useful during the design stage. They include: Weballey.net (numerous useful links), Activejump.com (good HTML tutorial), and Accessv.com (a good HTML tutorial can be found at http://www.accessv.com/~email/webpages/).

✪ ALL-IN-ONE WEBSITE SERVICES

Today you can find numerous web-based all-in-one e-commerce solutions. These web-based services provide server space and the ability to set up and run an online store through the simple process of inserting the proper information into web-based forms and using a point-and-click interface. These services are easy-to-use because everything is templated. But templates mean that you forego the unlimited flexibility of a customized program, such as personalizing your website to present a custom look to each visitor. Another thing about all-in-one services — few can handle large downloads. So if your business model dictates that you sell or offer downloadable products (software, music, video, etc.) there are a few items you should confirm before signing on the dotted line. They include: Will the service let your website serve large downloads? How much will it cost (e.g. will you be charged extra if your traffic exceeds a certain milestone)? Can their facilities handle the downloads (do they have a big enough pipe to handle your downloads and their other traffic)?

Still, for many small e-commerce businesses, the all-in-one solution is a good fit. Especially since these services allow you to take advantage of industry-leading e-commerce design power without the necessity of purchasing or maintaining sophisticated e-commerce software. But if you envision your e-commerce site "standing out in a crowd," this is not the solution for you.

If you are interested in taking the all-in-one service route, go with a well established service. The best known of the all-in-one website services is perhaps Yahoo! Store, followed by (in no specific order) Terra Lycos' Angelfire.com, eBizWebpages.com, Digital River's FreeMerchant.com, Merchandizer.com, and Bigstep.com. While the fees for such services range from $5 to $300 per month, it is possible to find services that offer basic store building tools for free, e.g. Angelfire.com. However, there is a catch — you must allow the service to place advertising on your site.

All of the previously mentioned services offer nice, robust store building tools that are easy to use, even for the novice. Yet, the actual services offered vary widely, even among the established all-in-one services. For example, some offer features such as a universal shopping cart with a single check-out no matter how many stores are visited, one-click buying, an easy-to-browse product catalog, and product recommendation and comparison engines. Others may offer individual shopping carts, but help with establishing a merchant's account, and/or real-time credit card processing. You also will find all-in-one services that offer few e-commerce specific tools except a shopping cart (e.g. no shipping options, accounting export, comparison engine, etc.).

These all-in-one services, however, also tend to offer "extras" that can help a fledging e-commerce website expand its business. But the services offered — surveys, FAQs, press releases, maps, registration of your site with search engines, etc. — differ from service to service.

All-in-one web services may be just what you are looking for to ease your entry into the world of e-commerce, if so check them out; one may fit your needs. Even larger businesses may find that one of these services can suit their needs — in the short term. For example, to have a good, viable store that could meet the expectations of its 1999 Christ-

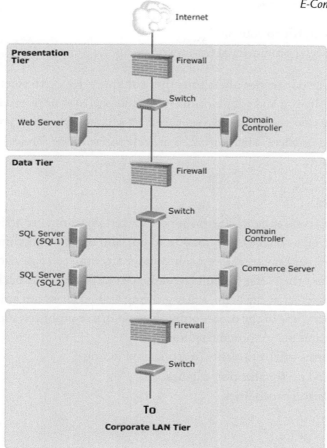

Figure 22. A simplistic diagram of a website built around a commerce server.

mas shoppers, FAO Schwarz (a very large toy store) placed its catalog in Yahoo! Stores. Be aware though, that many all-in-ones have very restricted space for catalog inventory.

While all-in-one services offer many benefits, you should understand that these services, out of necessity, limit the breadth and scope of a business's growth.

Whatever solution you are considering, be sure to get answers to the following questions before making your final decision.

* Can I use my domain name?
* What type of credit card processing is offered and is there an additional charge for this service?
* Am I required to host ads and/or other types of branding?
* What is free and what do I have to pay for? Read the fine print and get down to specifics or you might find yourself with an unpleasant surprise at the end of the month.

* How will they help drive traffic to your site?
* Does the service offer special promotional features?

Remember that each all-in-one service offers something unique — it's up to you to decide which solution best fits you and your e-commerce business. Do your research, and then perform the necessary due diligence to ensure that whatever company you choose is stable and has the financial wherewithal to provide you with the services it promises year in and year out.

✪ COMMERCE SERVERS

Commerce Servers are web servers enhanced with support for certain commerce activities. Commerce server platforms can be costly ($800 and up), and most are anything but plug-and-play. However, they can be the best solution for any mid- to large-sized e-commerce business because they offer the full e-commerce platform in one package. By placing a commerce server at the helm, a fast-growing web business can ameliorate or sidestep altogether some of the more common e-commerce problems such as performance, scalability and integration difficulties.

Commerce server platforms offer a diversity of services including:

* Transaction, payment and personalization engines.
* Tax and currency calculation capabilities.
* Workflow automation.
* Content management software.
* Database and ERP (enterprise resource planning) integration modules.
* Proprietary and open application servers.
* Customer-service offerings with "800" number phone support integration.
* Smart catalogs.

Although commerce servers have come a long way in the last few years, some still lack support for necessities such as back orders or contracts that require purchases at specific intervals. Also proprietary architectures often make it difficult to customize some commerce offerings, and back-end integration issues may require writing your own custom code for each legacy system that needs integration.

Most of the products don't have the modularity necessary to meet the performance, scalability, and integration needs of today's web-based businesses. However, that being said, unless you have the resources to write your own software, you will be forced to buy into a commerce server solution, whether or not the technology is "up to snuff."

That is the bad news, now for some good news. There are some good commerce servers on the market. Here is a taste of what today's marketplace has to offer in the way of innovative commerce servers.

Advansis E-Commerce Server (www.advansis.com). This turnkey solution features all the components needed to launch a comprehensive online store, including a browser-based product catalog manager, shopping cart, online ordering, and extensive payment, security, and order processing features. Furthermore, the product can be easily scaled to support an unlimited number of products. It also provides an affiliate program manager and customizable currency options. On top of all that, the Advansis E-Commerce Server includes support for custom catalog design using templates, stock monitoring capabilities, "quick search" capabilities, pre-configured payment provider options, and support for an unlimited series of tax and shipping schedules.

BEA WebLogic Commerce Server (www.bea.com). Designed in Java and based on open standards, the BEA WebLogic Commerce Server blends ready-to-run, out-of-the-box commerce functionality with the scalability and flexibility needed for e-business operations. It also simplifies development of portable and scalable applications, and provides interoperability with other applications and systems. Moreover, the latest version of the WebLogic Commerce Server includes the WebLogic Personalization Server with Portal Framework, which enables a web business to create adaptable e-commerce applications that personalize customer interactions.

Chapter Eight's Commerce Server 2003 (www.chaptereight.com). This product enables an e-business to safely transact business in a secure, robust, and a cost effective manner. Its key features include the ability to conduct credit card transactions safely, to track orders via an order management system (or link to an existing order management system), to analysis orders so you can discern sale patterns, and the ability for customers to manage their orders online.

IBM Corp.'s Websphere Commerce Suite (www.ibm.com). This product is based on an open Java application technology, which the large enterprise may find very useful. The Websphere Commerce Suite provides a viable infrastructure for building, deploying, and administering e-commerce sites. IBM also has done a nice job in making enterprise-level provisions and tools (e.g. support for auctions, business intelligence, personalization, and marketing campaigns) sophisticated enough to address the needs of a maturing e-commerce site. If you envision your business providing global commerce, then the Websphere Commerce Suite might be a good fit. It is designed to enable an e-business to craft its site so that it can be customized to meet requirements in foreign markets. Parameters for language, currency, and taxes (including value-added taxes) are easy to select. Other "global" features include regional shipping guidelines and carrier selection, and currency selection. You can even provide context-rich page displays relevant to a specific shopper's point of origin.

Microsoft's Commerce Server 2002 (www.Microsoft.com). This server is a viable option for the small to mid-size website. In true Microsoft style, its Commerce Server 2002 doesn't exactly mimic the model used by its competition. Where IBM, BEA, Netscape and others tend to create an application server upon which e-commerce modules are added, Microsoft's approach is more monolithic. Commerce Server contains features for personalization, business-decision support, content management, order processing, and much more. Just about the only thing not included in the package is an XML engine (that can be found in Microsoft's BizTalk Server). If you would like to know more about Microsoft's approach to commerce servers go to www.microsoft.com/technet/default.asp, the select "Products and Technologies" (menu — left side).

SeeCommerce's Dynamic Commerce Server (www.seecommerce.com). This Java-based product manages and distributes information (structured and unstructured) from a variety of sources, e.g. data marts, data warehouses, web servers, ERP systems, operational databases and corporate legacy systems. As such, it enables an e-business quickly and securely to design, to generate, to publish, and to distribute information to its customers.

Simple Logic's SecureMerchant e-Commerce Server (www.simplelogic.com). This dynamic application foundation is ideal for online retail catalogs. The SecureMerchant e-Commerce Server combines the latest technologies, features, and security needed for today's e-business environment. Its web-based merchant administration features allow store management from any location and the vendor's comprehensive online help services can reduce ramp-up time. Most importantly, businesses enjoy increased sales because customers can find what they want easily.

When looking for commerce server software, look at products that provide:

* A modular system (to enable you to choose the best fitting elements for your enterprise).
* Rules-based workflow automation, especially for content management.
* Support for monthly supply replenishment or back-order provisioning.
* EJB/XML support.
* Scalability when the shift from back-end systems to multi-company workflow integration (i.e., product descriptions that will be provided directly by the supplier) becomes the norm.

There is, in all probability, a commerce server platform out there suitable for any particular web-based business model. However, if you are considering building your site upon a commerce server, choose carefully. Then be sure that your final choice can integrate seamlessly into your legacy systems. Also, consider your budget, existing architecture, and software development capabilities when making the decision to purchase a commerce server platform.

✪ CONCLUSION

This chapter educates the reader on the technology needed to turn a website into an "e-commerce business." But it is up to you and your e-commerce team to choose the method by which you implement that technology.

Use your blueprint as your guide when selecting any e-commerce solution. Don't rush your decision-making process —take the time to determine how you want to implement the actual "e-commerce" features into your website.

Finally, never lose sight of the fact that your chosen e-commerce solution(s) must work seamlessly with not only your hardware and other software, but also with your means of collecting payment. For instance, if you are to process credit card payments online, the chosen e-commerce solution must offer total payment processing integration — payment, gateway, and the agencies involved (i.e. the banks making and accepting fund transfers, and the relevant credit card processing companies).

Chapter 10

Adjunct Software

Build momentum by accumulating small successes.

—Anonymous

Up until this point we've discussed what is needed to build a basic online presence. But many e-commerce businesses will want more than just the essentials. The three most popular e-commerce add-ons are online auctions, file sharing (peer-to-peer networking), and weblogs. Here is how each of these models works.

✪ ONLINE AUCTION

Online auctions are popular. People like the thrill of the chase, the satisfaction of capturing their quarry, and the pleasure of achieving a bargain (or a profit).

An online auction can be a dynamic and profitable business. And almost any type of merchandise can be sold via an auction format, whether you're liquidating inventory, selling memorabilia, antiques, books, computers and their peripherals, or just bringing buyers and sellers together.

There are a couple of popular online auction models. The first is where the e-business owner physically controls the product being sold. As such, the e-commerce site accepts all payments for the goods. Successful bidders will usually use the same payment methods as used on the typical e-commerce site: credit card, debit card, personal check, cashier's check, money order, or escrow service. There are many good examples of this type of auction site including CollectorAuctions.com, Biddingtons.com, and Southebys.com.

The other model is a website that enables person-to-person auction activity. eBay

AUCTION FORMATS

Some of the more popular online auction formats include the following. But note that although these five formats operate as set out herein, they do not always carry the same moniker as used in this list.

English or Regular Auction — With this auction format, the seller can start the bidding at a certain price and have each bid be incremented at a set rate. For example, if the seller had a camera that he or she wanted to sell for around $50, that seller would state a starting bid of $40, and then have each additional bid go up by $5. So the first person to bid would have to bid $40 and the second would bid $45 and the third would bid $50 and so on.

Reserve Price Auction — This auction format allows the seller to set a minimum price for which he/she is willing to sell the item. This "lowest price" is called the "reserve price." In a reserve price auction the bidders know there's a reserve price, but not the actual price reserved. In order to win the auction, a bidder must 1) meet or exceed the reserve price, and 2) have the highest bid. If no bidders meet the reserve price, the seller is under no obligation to sell the item to the highest bidder.

Private Auction — Strictly hush, hush, this auction format protects a buyer's privacy. Unlike other auction formats, bidders' email addresses will not show up on the item or bidding history screens. When the auction is over, only the seller and the highest bidder know who bought the item.

Dutch Auction — This auction format is often times used when the seller has many identical items to sell. With the Dutch auction format, the seller 1) lists a minimum price (or starting bid) for one item, and 2) the number of items for sale. Bidders specify both a bid price and the quantity they want to buy. All winning bidders pay the same price per item — which is the lowest successful bid (which might actually be less than what the bidder bid). If there are more buyers than items, the earliest successful bids get the goods. Also, higher bidders are more likely to get the quantities they've requested. However, bidders can refuse partial quantities. For example, if you place a bid for ten items and only seven are available after the auction, you don't have to complete the purchase. Here are some examples of how Dutch Auctions work:

is the most popular of this type of online auction business model. Individual sellers and/or small businesses auction their items directly to the consumer. With this model, the seller — not the website owner — has responsibility for the merchandise, payment for said merchandise, and shipment to the purchaser.

This type of auction site usually requires that sellers register, obtain a "user account name" or "screen name", and meet other criteria, before they can place items for bid. Sellers also must agree to pay a fee every time they conduct an auction. Most sellers set a time limit on bidding and, in some cases, a "reserve price" (the lowest price they will accept for an item). When the bidding closes, the highest bidder "wins." If no one bids at or above the reserve price, the auction closes without a "winner."

At the end of a successful person-to-person auction, the buyer and seller commu-

A seller has ten new Seagate hard drives for auction at $10 each. Ten people bid $10 for one hard drive each. In this case, all ten bidders will win a hard drive for $10. But, if five people bid $12 for one hard driven each and seven others bid $10, the minimum bid for the hard drives is still $12, but because the five $12 bidders bid higher than the $10 bidders, they are guaranteed a hard drive. The other five hard drives will go to the earliest $10 bidders. Thus, with Dutch auctions, the final price for each an item may be lower than the highest bid placed, since all winning bidders pay the same price — which is the lowest successful bid.

Dutch Auctions might sound complicated, but the majority are simple because most users win the items they bid on at the minimum asking price. Still, there are some special instances you might want to know about:

If you are the lowest bidder in a Dutch Auction and you specify a multiple quantity, you may not get to purchase all that you specify. Why? Because there may be little left over after the high bidders get their share.

In other words, if the lowest bidder requests a quantity of four hard drives, he or she may only get one since the first nine drives have already been allotted to the higher bidders. The only way to avoid this problem is to make sure you are not the lowest bidder.

In Dutch Auctions (multiple item auctions), successful high bids are usually displayed via a "High bidders" link. And the complete bidding history (including any unsuccessful bids) is typically displayed via a "Bid history" link.

Restricted Access Auction — This auction category is normally used for "adult" material such as erotica. By using the Restricted Access Auction, customers can easily find — or avoid — "mature audience" merchandise. This allows customers to avoid inadvertently viewing explicit materials.

Note, however, that sellers of mature-audience-items may be subject to various legal statutes that regulate the sale and distribution of mature-audience-related materials. And if you plan to allow your website to offer "adult" items for auction, you should place a notice to bidders and buyers of adult items that informs then that there are various legal statutes that regulate the sale and distribution of those materials.

nicate — usually by email — to arrange for payment and delivery. Yahoo! Auctions, eBay, and Ubid are the most popular person-to-person auction sites. But also see the "Auction" section of Chapter 1 for examples of smaller, niche websites that represent this model.

Requirements

In order to set up an online auction site, whether as an adjunct to an existing e-commerce site or as a standalone business, you typically need:

* A web server that supports Perl 5 or above and CGI scripts.
* Auction software and specialized script.
* Some hard disk space on the web server for the script to write auction data. Usually

one MB of space will do when starting out, but more may be necessary if your auction becomes popular.

✳ Time and patience. A typical installation of auction software only takes a few minutes. But those of you who want your auction site to look original or who are attaching an auction module to an existing e-commerce site, you might want more. For instance, you may need to tweak the script a bit to give your auction pages an original look and feel, or in order to match the auction pages to rest of your website.

Software

Here is a list, not inclusive, of auction software to consider using to setup your auction site.

Auction Bid Software (www.biddotcom.com). This is one of the more popular auction software products, probably because no special skills are required to get it up and running. And updating your site is as simple as typing an email message; putting those changes online just requires a click of your mouse. Cost — $299.

Auction Software for Online Auction (www.mewsoft.com/Products/). This auction software is driven by a SQL database and provides a variety of features including unlimited nested categories; multi-lingual capabilities; proxy bidding system; featured category, home page and gallery auctions; and Dutch, reserve, and private auctions auction types are supported. It also has a user accounts manager and advanced search tools. Cost — Free.

AuctionWeaver Lite (www.siteinteractive.com). This is a CGI program written in Perl that enables a website to create and host auctions. Its key features include unlimited categories and items; item preview before posting; outbid emails automatically sent (which means you also need access to an SMTP mail server); integrated search engine; and one script file for easy configuration and use. Cost — Free.

BidFlux Auction Software (www.bidflux.com). This fully functional online auction software is 100% self-contained. It is designed to be user and customer friendly, as such it has many similarities with the auction program used by eBay. BidFlux also allows for flexibility and growth as your auction business expands. Cost — $300.

E-Z Auction 1.8 (www.e-zauctionsoftware.com). This affordable auction software is very popular with the e-commerce crowd. That is because E-Z Auction 1.8 allows you either to create a standalone auction website or add auctions and classified ads to an existing e-commerce site. Some of its main features include: required registration, email verification, password protection, automated email notification, automated closing, integrated search engine, bidding history, browse closed auctions, change registration info, seller sets starting bid, seller sets reserve price, automated user statistics, unlimited subcategories, customized email, buy-it button, shipping and payment options, feedback, lost password, forgot alias, view sellers other auctions, and much more (e.g. ability to view all new, hot, and closing auctions). The program is also fully customizable and comes with complete resell rights. Cost — $10.

Emaze Auction (www.emaze.com/auction.cfm). Although a bit pricey, Emaze Auction 2.5, written in Allaire Cold Fusion, is easy to install, to manage, and to modify for any website's design needs. This auction software easily integrates into almost any existing e-commerce site or it can be used as the starting point for a new auction site. It is completely web-based, incredibly powerful, easily expandable, and fully customizable. A few other salient features include simple and fast installation, procure/classified/seller add-ons, user friendly administrative screens, ability to uses MS Access or MS SQL databases, CyberCash and AuthorizeNet ready, and its source code is readily available. E-businesses worldwide use Emaze Auction 2.5 and it has been translated into German, French, Swedish, and Spanish. The software is utilized to auction anything from personal products to autographs to estate jewelry to optical equipment to oil field pipe. Cost — $995.

Ultimate Auction (www.ultimate-auction.net). This auction software is written in perl and it utilizes the mySQL database (a binary database that can hold virtually unlimited amount of information without a change in speed). Ultimate Action is written so the average e-commerce operator can handle extensive online traffic and business needs easily. The product is also perfect for the small business that wants to start a small Dutch auction on their website. A few of its many features include the ability to create cookies, blacklists, and watchlists. It also offers good account management features, web-based administration, proxy bidding, a nice clean layout that loads quickly under most conditions, and encrypted passwords. Cost — $300 without installation ($400 with installation).

As some readers may have ascertained, much of the auction software discussed in this section is little more than what an experienced programmer could create with a web design package such as Macromedia's Dreamweaver MX. Though, of course, it would help if the programmer has experience in developing online auctions so that he or she could understand how to design the workflow for any number of pages dealing with the intricacies of buying and selling online. Nonetheless, using an auction package is probably the better choice since most e-commerce start-ups are not programmers and are on a tight budget. Even if you have an ample budget you may find one of the more costly auction programs to be a better choice than letting an experienced programmer have a go at designing your auction pages — building a bespoke system can run into a wad of money. Still, if you want an auction site that offers all the "bells and whistles," you may need a programmer.

Tips

Know Your Legal Obligations. Under federal law, you're required to advertise your product or service and terms of sale honestly and accurately. You can't place "shill" bids on an item to boost the price or offer false testimonials about yourself in the comment section of Internet auction sites. You're also prohibited from auctioning illegal goods. If running a person-to-person auction site, put in place a system whereby you can monitor the merchandise offered for auction to ensure that illegal items are not offered — the responsibility for ensuring that a sale is legal rests with the auction site owner, the seller, and the buyer. Consider posting a list of prohibited items as a guide for both buyers and sellers.

If operating an auction to market your own merchandise know that you must follow the following rules:

Delivery of Purchased Goods. You are required to ship merchandise within the time frame specified during the auction, or, if a time frame is not specified, within 30 days. If you can't meet the shipping commitment, you must give the buyer an opportunity to cancel the order for a full refund or to agree to a new shipping date. To learn more about your responsibilities when shipping products, see "A Business Guide to the Federal Trade Commission's Mail or Telephone Order Merchandise Rule," which can be found at www.ftc.gov/bcp/conline/pubs/buspubs/mailorder.htm.

Advertising Products. When describing an item and its condition, you must state whether it's new, used, or reconditioned. Specify the minimum bid at the lowest fair price you're willing to accept and who is to pay for shipping (also whether or not you are willing to ship internationally). State your return policy, including who's responsible for paying for shipping costs or restocking fees if the item is returned. Try to anticipate questions buyers might have and to address them in the description of your item or service. If possible, include a photo of the item. Don't forget to advise prospective bidders whether you provide follow-up service — or tell them where they can get it.

Dealing with Bidders. Respond as quickly as possible to bidders' questions about your item or your sales terms. When the auction closes, print all information about the transaction, including the buyer's identification; a description of the item; and the date, time and price of the bid. Save a copy of every email you send and receive from the auction site and/or the successful bidder. Contact the "winning" bidder as soon after the auction closes as possible; confirm the final cost, including shipping charges, and tell the buyer where to send payment.

Arranging for Payment. If you accept direct credit card payments from the buyer, bill the credit card account as close to the ship date as possible. If a buyer insists on using a particular escrow or online payment service that you've never heard of, check it out — visit the service's website, call the customer service line. If there isn't one, or if you call and can't reach someone, don't use the service. Also, before agreeing to use an online payment or escrow service, read the terms of agreement:

* If it's an online payment service, find out who pays for credit card charge backs or transaction reversal requests if the buyer seeks them.

* Examine the service's privacy policy and security measures. Never disclose financial or personal information unless you know why it's being collected, how it will be used, and how it will be safeguarded.

* Be suspicious of an online escrow service that cannot process its own transactions and requires you to set up accounts with online payment services. Legitimate escrow services never do this.

* Check with the Better Business Bureau, state attorney general and/or consumer protection agency, both where you live and where the online payment or escrow service is based, to see whether there are any unresolved complaints against the service. Be mindful that a lack of complaints doesn't necessarily mean that the service has no problems.

✪ PEER-TO-PEER

A peer-to-peer (P2P) operation is where individuals within a niche market open their computers and allow others to upload files directly from their hard drives. Sharing music, video, and software are the most popular reasons for adopting a P2P business model, but there are many ways a P2P model can be used to enable a niche market to share information and files. For example, film students could share their creations, bird lovers could share photos, migratory information, etc., and software programmers find P2P networks invaluable when writing code.

There are a few different types of P2P networks. Let's examine the most popular models.

Napster model. This refers to a centralized version of peer-to-peer file sharing that is designed around a central server that directs traffic between registered users. The central server maintains a directory of all registered user's shared files (which are stored

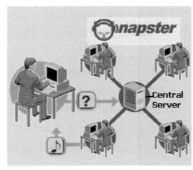

Figure 23. The difference between a Napster-like network and a Gnutella-based network.

on each individual's computer). To find a specific file, a registered user types in search criteria (e.g. name of song, artist, book title, genre, etc.) and starts a search of the system. The central server then retrieves a list of files matching the search request, which it presents to the requesting user. The user then selects the desired file(s) from the list and opens a direct link to the individual computer that contains the desired file(s) and downloads the file(s) directly from the other user's computer. Actual files are never stored on the central server. The problem with this version of P2P networking is that if the central server goes down the entire network is unavailable to end-users.

Gnutella model or decentralized network. This model originated from the Gnutella protocol written by Justin Frankel, the 21-year old founder of Nullsoft. It can be described as a type of call-and-response protocol, i.e. a more complicated email or news protocol.

Nullsoft posted Gnutella on the Web in March 1999. Then in June 1999 Nullsoft was acquired by America Online (AOL), and a day later AOL yanked Gnutella, but several thousand downloads had already occurred, allowing reverse engineering to take its course. Thus Gnutella continues today, albeit under the auspices of independent programmers who offer popular P2P software packages such as LimeWire, Kazaa and BearShare. These software packages and many more can access the same "Gnutella" P2P network — a network that does not use a central server to keep track of users' files. Each instance of a Gnutella-type software acts as *both* a server and a client because the software supports bidirectional information transfer.

All you need to operate a fully functional Gnutella network is to install any of several available Gnutella software packages (Kazza, BearShare, etc.) and to find other Gnutella clients that are willing to communicate with you. You will communicate directly only with the handful of sites you've agreed to contact. Any material of interest to other sites will pass along from one site to another in store-and-forward fashion. Since Gnutella networks use the Internet, if your computer is running Gnutella-based software, you can connect with someone who's in another country just as easily as within your own region. This means peer-to-peer networks are robust and virtually failsafe.

Under the Gnutella model, the user starts with a networked computer equipped with a desktop version of Gnutella (e.g. LimeWire, Kazaa, BearShare), which is then used in conjunction with an ISP account to connect to the Gnutella network and to announce to all other computers connected to the P2P network that it is alive. The user's computer can then search the contents of the shared directories of the P2P network, and download the desired files directly from another user's computer.

The advantage of using this type of P2P network is that it does not rely upon a central server, thus it would be virtually impossible to bring it down. Even if distribution of P2P software were banned, the network itself could still continue with all the current users that have already downloaded the client software onto their computers.

There are different versions of the Gnutella model. Grokster and others have taken the Gnutella model and modified it. For instance, rather than all networked computers passing along each search query, the Grokster distributed self-organizing network recognizes faster computers that utilize a powerful connection (e.g. T-1/E-1) and selects those computers to be what is referred to as "SuperNodes," which then serve as local search hubs. When a non SuperNode computer connects to the network, its "neighborhood SuperNode" uploads information about that computer's shared files along with performing searches. The selection of a SuperNode is automatic, there is no manual intervention.

Freenet model. This P2P model is a distributed anonymous information storage and retrieval system. The best way to explain the Free Network Project (i.e. Freenet) is to quote the website's explanation:

✳ Freenet is free software that lets you publish and obtain information on the Internet without fear of censorship. To achieve this freedom, the network is entirely decentralized and publishers and consumers of information are anonymous. Without anonymity there can never be true freedom of speech, and without decentralization the network will be vulnerable to attack.

* Communications by Freenet nodes are encrypted and are "routed-through" other nodes to make it extremely difficult to determine who is requesting the information and what its content is.

* Users contribute to the network by giving bandwidth and a portion of their hard drive (called the "data store") for storing files. Unlike other peer-to-peer file sharing networks, Freenet does not let the user control what is stored in the data store. Instead, files are kept or deleted depending on how popular they are, with the least popular being discarded to make way for newer or more popular content. Files in the data store are encrypted to reduce the likelihood of prosecution by persons wishing to censor Freenet content.

* The network can be used in a number of different ways and isn't restricted to just sharing files like other peer-to-peer networks. It acts more like an Internet within an Internet. For example Freenet can be used for:
  - Publishing websites or 'freesites'
  - Communicating via message boards
  - Content distribution

Freenet is not just theoretical, it has been downloaded by over 1.2 million users since the project started, and it is used for the distribution of censored information all over the world, including countries such as China and the Middle East. Ideas and concepts pioneered in Freenet have inspired hundreds of academic papers in the fields of computer communication, security, and law. Freenet has also received significant coverage in the mainstream press.

Note: If you would like a more detailed tutorial on how peer-to-peer networks work visit the howstuffworks website (www.howstuffworks.com/ file-sharing3.htm.) If you would like more information on Gnutella and Freenet's inner workings there is a very good article by Andy Oram, which can be found at www.oreillynet.com/pub/a/network/2000/05/12/magazine/gnutella.html. Also visit http://freenet.source-forge.net/ for more information on Freenet.

Yaga model. This is a relatively new P2P technology. Yaga.com has crafted technology that provides built-in security (e.g. files are scanned for viruses before being made available), support for digital rights management, reliable transfer of gigabyte-size files (an increasingly important issue when distributing software and videos), and centralized management (files are given unique signatures based on their size). As soon as a file is made available, it is included in the search index, supporting full-text search (every word, not just the title and description). All the basic elements are there to enable content producers to use the Yaga network as a mechanism for selling content.

For example, with Yaga you can make your files available to a community of users, and also download files made public by others. Budding programmers who have written software that they would like to make available to the world would find Yaga useful. An up and coming rock band that is having trouble getting their music played on the radio can, with the help of Yaga software, make their product readily available to a community of users worldwide.

Yaga may be the best tool if you want to set up your own private branded P2P network. You could then use that network to publish software, a musical library, your latest video concept, photographs, books and manuals, or any other kinds of files — for free or for a price, depending upon your business model.

Content-Addressable Network (CAN) model. We can't conclude this discussion without addressing the experimental, but very scalable CAN model. Freenet's aim is to provide anonymity during both file sharing and lookups, and is intended to provide robustness in the face of hostile agents both (legally and technically); whereas, Gnutella was designed to provide a complete solution to file sharing in the post-Napster era. Both P2P systems are designed for immediate deployment. However, CAN is a building block for not only file sharing but also for higher-level services, such as DNS, distributed file systems, etc. While at this writing CAN is still largely a research rather than a production system, keep an eye on its progress. It offers some exciting possibilities. Go to your favorite search engine and perform a search to track the latest developments.

Requirements

To establish a P2P e-commerce operation by becoming a partner or an affiliate of an established P2P network, you need at minimum:

* A niche market group who will open their computers and allow others to upload files directly from their drives.
* A working e-commerce website, of course.
* P2P client software, which you would provide to end-users so they can access the P2P network. Then your targeted market will need to download that software and put their shared files into a folder on their own hard disk.

You can create your own members-only network or become a part of a larger P2P network such as Grokster, Overnet, eDonkey2000 network, kaZaa, or Blubster. It is expected that only a small minority of readers will want to operate their own P2P network — most will want to piggyback onto an already existing network. Thus, the design and maintenance of a P2P network will not be discussed at length in this book. But to point any intrepid soon-to-be P2P operators in the right direction, Peeracy.com (http://peeracy.com) provides source codes (most are free) for various P2P networks. Once you call up the home page, just click on "P2P Software" link. The first downloads listed are for P2P clients. Scroll through those and you will find a number of different source code downloads. And visit the Afternapster.com website. This website not only keeps an up-to-date list of the latest P2P networks and client software, it provides a good, but brief overall review of each product/website.

Read books on P2P networking. For instance, the author learned quite a bit about P2P from reading the *Peer to Peer: Harnessing the Power of Disruptive Technologies,* Andy Oram (O'Reilly). Other good books include *Peer-to-Peer Computing: Technologies for Sharing and Collaborating on the Net,* David Barkai (Intel Press), and *Peer-to-Peer: Building Secure, Scalable, and Manageable Networks,* Dana Moore, John Hebeler (McGraw-Hill Osborne Media).

Business Model

Wondering what P2P has to offer an e-commerce operation? There are ways to make money with P2P, although it is doubtful that you will ever get rich from such an operation. For example, you could —

* Enter into contracts with third party software providers (e.g. Gain, Mysearch, Shopathome, New.net, etc.) who will pay you for distribution of digitally rights-managed files.
* Sell advertising space on the client software that is downloaded onto your customer's computers.
* Sell a version of the client software that is ad-free.
* Sell products and service that is of interest to your niche market.
* Create content, post that content on your website to encourage people to come to your site, and then sell banner and pop-up advertising space on your website.
* Create a newsletter and sell sponsorships and/or ads.

Media	56 Kbps Modem	Cable (average 512 Kbps)	T-1/E-1 (1.54/ 2 Mbps)
Photo, 200 KB	40 seconds	2 seconds	2 seconds
MP3 Music track	13 min 30 seconds	1 minutes	15 seconds
Full-length movie	22 hours	1 hour 45 minutes	25 minutes
Five-minute video clip	1 hour	6 minutes	2 minutes
Novel, 1 MB	4 minutes	15 seconds	4 seconds

Table 6. The average download speed for some of the more common files shared over a P2P network.

Tips

Your P2P network doesn't necessarily need to serve music or video enthusiasts. You could use your P2P e-commerce site to enable a community of content developers to share files, or perhaps cooks to share recipes, travelers to share travel journals, cartoonists could share their creations, and programmers also would find P2P a valuable tool. Another idea for using the P2P model is to enable individual stock traders to share court filings, SEC documents, news articles, etc. about various companies — worldwide.

There are many ways a P2P model could be used to enable a niche community to publish their endeavors and easily to share it with a global audience.

✪ WEBLOGS

Weblogs are the newest Internet craze — there are an estimated 500,000 active weblogs available to the surfing public. Commonly known as "blogs," weblogs are so popular that the May 2002 issue of *Wired Magazine* predicted that "Weblogs will outrank the New York Times website by 2007 (based on a Google search of five keywords or phrases reflecting the top five news stories)."

A weblog is a web-based writing space or online journal. And, like a journal, a weblog is a dynamic, continually updated website that grows over time with the accumulation of writing and other content. Typically, weblogging (or blogging) takes place within a community of other webloggers (bloggers) who share a common interest.

One of the reasons weblogs have spread so rapidly is their ease of use — the writing, editing, look, and feel of the site can be managed through your web browser. A good weblog is designed so that a new entry can be "turned" each day (just like a journal). In addition, a weblog site should be able to keep track of the date, and archive all entries.

Weblogs are ideally suited to interaction between people sharing special interests. Thus, many e-commerce businesses may want to consider adding a weblog to their website. The weblog's capacity for information dissemination and feedback potential can tap into the buying power of a blogging community.

Requirements

Your weblog requirements will vary depending on what type of weblog you want. Is it to be added as an adjunct to an e-commerce website or to stand by itself? Is it to be self-hosted or to be hosted by a weblog service?

If self-hosted or part of an existing e-commerce site then the infrastructure requirements are the same as for any other type of e-commerce website, except for the addition of XML (actually RSS, a dialect of XML) and a database (if one is not already being utilized). Weblogs also need a set navigation structure and a means to archive the daily entries automatically.

If you go the weblog service route, all you need is a computer with an Internet connection, a viable email address, and possibly access to a web server with FTP (File Transfer Protocol).

Software

We will now discuss some of the software and tools you might want to consider using to establish a weblog.

RSS

Before you implement a weblog, you must understand its underlying technology — the XML dialect referred to as "RSS." RSS provides the means for organizing and publishing the content on your web pages. The acronym "RSS," which can stand for either "really simply syndication" or "RFD site summary" (the differences will be explained later), refer to the code that underlies a weblog. Specifically, RSS (whatever version) is a protocol that is an application of XML, which provides an open method of syndicating and aggregating web content, including weblogs.

Through the use of RSS, you can create a data feed that supplies headlines, links, and article summaries from your weblog to your niche market (as long as they ask for the feed). This allows interested parties (e.g. your niche market) to obtain constantly updated content from your website through the help of a news aggregator.

As mentioned previously, there are two types of RSS currently used as the underlying code for weblogs.

RSS 0.91 is referred to as "Rich Site Summary." Netscape released the original popular version 0.91 in July 1999. That initial RSS code has since been upgraded by Dave Winer of Userland — first to 0.92, then 0.94, and now 2.0.

RSS 1.0 is referred to as "RDF (Resource Description Framework) Site Summary." This version of RSS provides an XML structure for describing document metadata content. The RSS-DEV Working Group created this RSS version to support the Resource Description Framework. RSS 1.0 documents can draw upon any RDF-compatible extension syntaxes called modules.

RSS Editors

Don't want to worry about the RSS coding or adding the cost of an authoring/publishing application to your budget? If so, you have a number of options. While you can create a RSS file with any type of text editor, in doing so you run a real risk of syntax errors. Thankfully, there are plenty of free RSS-specific web-base tools for your use. Make use of one of the following free web-based RSS editors:

* Ukoln's RSS-Xpress editor forms RSS 1.0 XML. Go to the Ukoln's website (http://rssxpress.ukoln.ac.uk) for more information on this tool.
* WebReference's RSS Channel Editor generates valid RSS 0.91 files. To access this tool go to the WebReference website (www.webreference.com/cgi-bin/perl/rssedit.pl).
* WebDevTips, which also offers the RSS Headline Generator, constructs RSS 0.91 feeds. It can be found at www.webdevtips.com/webdevtips/codegen/rss.shtml.

After a file is complete, save the files locally, and then load them to your server when applicable.

For more detailed information on RSS and authoring/publishing a weblog:

* Visit these two websites — http://blogs.law.harvard.edu/tech/rss and http://web.resource.org/rss/1.0/. Both sites discuss in detail the two incompatible versions of RSS: RSS 1.0 and RSS 0.9x/2.0.
* Use the RSS Workshop which is available on the State of Utah's website at http://gils.utah.gov/rss/.
* Read *Content Syndication with RSS*, Bill Hammersley, (O'Reilly). O'Reilly has placed on its website Chapter 4 of the book, which explains the Really Simple Versions of RSS (www.oreilly.com/catalog/consynrss/chapter/ch04.pdf). Another good book is *The Weblog Handbook: Practical Advice on Creating and Maintaining Your Blog*, Rebecca Blood, (Perseus Publishing).

Note: *All RSS files must conform to the XML 1.0 specification, as published on the World Wide Web Consortium (W3C) website.*

Authoring/Publishing Tools

To design your own full-blown weblog with capabilities similar to, for example, the Steve Gillmor weblog on CMP Media's CRN website (www.crn.com/weblogs/stevegillmor/), the Ken Layne and the Corvids weblog at www.kenlayne.com, the actor Wil Wheaton's weblog at www.wilwheaton.net, or the Nick Denton weblogs at both www.gizmodo.com and www.gawker.com, you will need authoring/publishing tools.

Weblogs are content-oriented with much more text than graphics, and most have some degree of collaborative editing capability, thus you can use a regular content management system both to create and to manage a weblog. (The only real distinction between a weblog and other types of web-based content is the RSS coding and the chronological element of weblogs.) Examples of software systems that are suitable for weblogs include the following:

Manila (http://manila.userland.com). This robust Internet server groupware allows a collection of writers, designers, and graphic designers to manage a website (including weblogs) through an easy-to-use browser interface. (Blogue.com uses Manila.) Cost — $900.

Movable Type (www.movabletype.org). After this powerful, customizable publishing system is installed on your web server, an individual or a group of collaborators can manage and update weblogs, journals, and site content. Costs — although Movable Type is free for personal or non-profit use, commercial users must pay a licensing fee of $150.

pMachine (www.pmachine.com). This innovative software enables the publication of virtually any kind of web content — from a basic weblog, to an advanced interactive magazine. pMachine gives you complete control over the presentation and the interactivity of your website. It is also simple to set up and use, allowing you to have your first weblog entry up-and-running in record time. (The Blogalization community used pMachine.) Cost — $125 for commercial license.

Scoop (http://scoop.kuro5hin.org). This collaborative media application falls somewhere between a content management system, a web bulletin board system, and a weblog tool. That is because Scoop enables a website to become a community. It empowers a site's visitors to be the producers of the site — contributing both news and participating in discussions. Cost — Free.

w.bloggar (http://wbloggar.com). Billed as "the best interface between you and your blog," this application is both a "post" and a "template" editor, and sports a number of features and resources that the average browser-based weblog editors (discussed later) don't offer. For instance, because w.bloggar runs over a Windows GUI it allows the writer to edit his or her posts offline, and then save one or more text files locally for posting on the weblog at a future time. The software stays minimized to the "tray bar" as an icon until the writer wants to write and/or publish new text. All that is needed is for the writer to click the icon and the editor is ready to go to work. Once the text is finished, another click (if connected to the Internet) and the post is published to the website. Out of all of the previously mentioned tools, w.bloggar is compatible with the most weblog systems (e.g. e.bloggar can post and publish to Blogger, b2, MovableType, Nucleus, BigBlogTool, BlogWorks XML Blogalia, and Drupal blogs). Cost — Free.

Zope+Plone (www.zope.org *or* www.plone.org). Zope+Plone is not your average

NEWS AGGREGATOR

Just as a web browser can "read" a page of HTML, a news aggregator (also known as news-reader) can "read" RSS. You may notice when you visit a weblog that some have a small orange XML or blue RSS pictograms displayed — both of these little icons indicate that the page is available in RSS, i.e. news aggregators can automatically collect updates. News aggregators read news, weblogs, and other types of "feeds" on the Web and aggregate them so the news feeds are readable from a single place, regardless of their source.

To subscribe, however, the reader needs a news aggregator client — a piece of software that decodes RSS feeds. There are a number of free news aggregator clients available including NewsGator (www.newsgator.com), which runs within Microsoft Outlook; RSSReader (www.rss-reader.com); NewzCrawler (www.newzcrawler.com), which gathers news content from a list of source channels and displays them in a customizable view (e.g. scrolling list, news balloon and news ticker); and Aggie (available on the SourceForge.net website).

blogging software. This expensive, robust, multiplatform system requires the programming language Python and you also need an experienced programmer on staff. The Zope+Plone system includes the Plone Content Management Framework [CMF] for portal construction and content syndication. Everything is stored in either Zope's built-in database or an external database application. Although its main use is as a content management system, Zope+Plone currently is used to support a number of heavy-duty weblogs including NATO, the Austrian Government, two U.S. governors, CBS New York, and Canada's Michael Smith Genome Sciences Centre, part of the BC Cancer Agency in Vancouver. Cost — varies according to website's needs.

Weblog Services

If you don't want to incur the expense or expend the time required to design your weblog site from scratch, you can use a weblog service. Many good weblogs are operated using such services, e.g. Amy Wohl (http://amywohl.weblogger.com/), Bag and Baggage (http://bgbg.blogspot.com/), Ernieattorney (http://ernieattorney.typepad.com/), Isen.com's blog (which is powered by Blogger).

Some of the more popular weblog services include Blogger.com, Blogspot.com, Typepad.com, and Weblogger.com.

Business Model

An e-commerce business can use blogging as a powerful, cost-effective tool for communicating with its customers. Some brochureware site owners are even considering replacing their current static pages with a dynamic weblog.

But, unless you are pairing a weblog with another e-commerce model, it may be difficult to generate income by writing a weblog. That being said, here are few ways to earn revenue from a weblog-based e-commerce site:

* Sell products and services that are of interest to your niche market.
* Enter into affiliate agreements with other websites, e.g. Amazon.
* Ask members of your blog community to make donations. If you craft your weblog in such a manner that its contents are perceived as valuable to your niche community, and that community is involved in the blogger's agenda, some will be happy to give a donation every once in a while. (That's why you see so many PayPal links on weblogs.)
* Syndication — but only if you already are well known in your niche marketplace (such as the previously mentioned Nick Denton).

Note: For more ideas of how to earn income from your weblog, you might want to purchase "Blog Profit Ideas Exposed" by Mani Sivasubramanian. To order a copy, go to www.blogprofits.com.

Tips

Until you are completely comfortable with posting your weblog's content, make use of one of the free weblog validation services. Despite its relatively simple nature, RSS is poorly implemented by many of the tools used by the average blogger.

Use a web-based validation service. These services can codify your text (i.e. translate it into code) to ensure you are producing RSS correctly. Two of the best free weblog validation services are:

RDF Validation Service (www.w3.org/RDF/Validator/), a service of the World Wide Web Consortium. All you need to do is enter a URI or paste an RDF/XML document into the site's text field box and a 3-tuple (triple) representation of the corresponding data model as well as an optional graphical visualization of the data model will be displayed.

RSS Validator (http://feeds.archive.org/validator/). This service only requires that you enter the address of your feed and click "Validate." If this validator finds problems in your feed, it will present a message indicating the problems and highlight the point where the problem first occurs in your feed.

Measure your weblog's popularity. Once your weblog is up and running you might want to use tools like Blogrolling's Top 100 and Technorati's Link Cosmos to measure its popularity.

Weblogs make a good collaborative tool. Consider using a weblog as a content management tool for internal collaboration. A number of companies (including Microsoft and Google) do exactly that.

✪ CONCLUSION

The longer visitors stay within the pages of your e-commerce site, the more familiar they become with it — familiarity breeds comfort. When a website visitor (your potential customer) feels comfortable they are more likely to trust whatever your website has to offer — products, services, opinions, etc. That trust will lead hopefully to a long-term customer relationship.

But to reach that point, it is essential that your website has a good deal of "stickiness." While good content and a pleasant design can help a website build a base of repeat customers, the *piece de resistance* is to create a sense of community within your e-commerce business. That is exactly what the software discussed in this chapter does — any or all of these tools can be used to help an e-commerce business build a community within its niche market. A vibrant online community translates into a sticky website; that, in turn, means more profit for the business owner.

Chapter 11

Quality Assurance

The first rule of intelligent tinkering is to save all the parts.

—Paul Ehrlich, in *Saturday Review*

A good Quality Assurance Plan (QA Plan) will determine how and in what order each aspect of your website should be tested. A QA Plan's first priority is to avoid a situation that will force you to accelerate or to change the plan hurriedly because of some unanticipated problem. For many websites, by the time a problem becomes too obvious to ignore, considerable financial damage may have already occurred. If you do not test often and thoroughly defects will accumulate or be missed. A good QA Plan will orient the project toward detecting defects early, close to the point of insertion, and not allow defects to infect work later on.

This can't be said enough — quality assurance testing is essential to developing a successful website. The ever-growing list of operating systems, software choices, browsers and user preferences, combined with the multitude of static web pages and dynamic database-driven websites, result in everything from customers having a problem-free experience while visiting your website to customers being driven crazy because their browser crashed. Most of the problems could be easily solved with a slight change in the HTML or other coding on the site.

✪ THE QA PLAN

To assure that visitors to your website have a pleasant and problem-free experience, develop and implement testing protocols to improve production specifications, visual and HTML style guidelines, and process flow — this is your Quality Assurance Plan.

A properly implemented QA Plan can:

* Solve browser incompatibility problems.
* Help you to stay current with HTML standards.
* Help you with the use of browser-specific HTML.
* Aid in the review of your site for bugs.
* Set up usability testing to insure that website user interfaces meet user needs.

Test every aspect — from Microsoft's Internet Explorer and Netscape to Opera and WebTV and other web appliances — for validation that your website provides a problem-free experience for your customers, no matter how they surf the Web.

After following the design guidelines set out in Chapter 2, and before your new website is ready for launch, "freeze" your prototype and intensively test it prior to giving the public access. Structure the tests per your QA Plan. Be prepared for fine-tuning and testing to go on for a month. You must stop all development and changes of the site during this testing period.

Consider investing in software tools that specifically test content accessibility, basic functionality, and behavior under controlled access loads. As your last test, perform stress testing to see how the entire system reacts under really heavy or "bursty" traffic.

Website testing tools can help illuminate what happens as traffic and load increase. However, some of the software that can help you implement your QA Plan can be hard to use. But persevere. Here are some products that might help you in your search for a trouble-free website (note that the cost of these products will depend upon your specific needs).

Watchfire's WebQA 2.0 Web Content Quality Testing tool (www.tetranetsoftware.com/products) helps put you in control of your website by supplying intelligence on simulated visitor interactions with site content and website transactions. Reporting, analysis, and measurement solutions provide you with a real-time view of your website. This vendor also offers, WebXACT, a free online service that lets you test single pages of web content for quality, accessibility, and privacy issues. You can access WebXACT at http://webxact.watchfire.com.

Webtrend's Enterprise Suite (www.webtrends.com) assists you in improving the quality, performance, and integrity of your website. It can illustrate broken links, chart biggest and slowest pages, document the loading time of connections, check the syntax of various HTML components, find the availability of external servers linked to your web server, and more, even crawling your website as a user would.

Segue's Silk Product family (www.segue.com) lets you test the performance of your website rigorously using as many simulated concurrent users as your site and network will support. Its ability to stress test web applications under heavy loads and simulate bursts of activity make it ideal for use by virtually any web-based business. In addition, its reporting features enable you to chart and to correlate response time results with server statistics to identify bottlenecks and problems quickly.

Site Mapper (http://trellian.com/mapper/) is perhaps the best of the lot. This product will analyze the content of your website and create a detailed map of all resources with an indexed listing by page and category. It also validates all links, so that visitors need never come across a "File not found" error. Some of the quality assurance features Site Mapper offers include: spellcheck of documents as they are mapped, built-in document preview functionality, the ability to integrate with your favorite document editor, it can map javascript files, stylesheets, and media files, and it can list and map external links and broken links.

Another quality service option is a full-service QA consulting firm like OTIVO (www.otivo.com). Services include browser compatibility testing, functional testing, and usability testing.

Testing the Prototype

The majority of your QA Plan should cover testing your website not only during development but also during your prototype stage before "going live." The optimal situation would be for your website to be developed on a development server and to be launched to a staging server for QA. We realize the smaller web-based businesses will not have the resources for these additional servers. Nonetheless, your QA Plan should set out a full testing plan to ensure you tested as much of the expected customer interaction and site functionality as possible.

Since your website's design should already meet the "browser neutral" criteria we specified in Chapter 2, you should test your design to see how it looks when viewed with:

* Different screen sizes.
* Different browser window settings.
* Different color resolutions.
* High security settings.
* Minimal feature settings.
* Different browsers and versions.
* Different client platforms.

When testing, manipulate the browser's options settings. Change them so that the page has a white background and the links are presented in the default color. Turn on the "don't load images" menu item, since some of your customers will have their browsers set this way. Note the results to each of the following questions.

* How does your website look with these settings?
* Can you still navigate your site with ease?
* Does your website use a text font that isn't one of the defaults? If so, does your site look okay?
* If you used Java on your website, test it with a browser that doesn't support Java as well as with a Java-enabled browser with that feature turned off.
* If your website requires a special plug-in, a special helper application, or special file type, test the website both with the special "whatever" installed and then without the special "whatever" installed.

If you can, test your site with WebTV and other web appliances. Some of these products are relatively primitive, generally having limited capabilities and lower resolutions. You'll find that web pages must be reformatted so they can be read on a TV screen.

Your QA Plan should take into account each static page and each of the unique templates so they can be tested at least once in each browser configuration to ensure cross-browser layout and functionality. All other non-browser-specific functionality should be tested in at least one to three browser configurations.

Next, your QA Plan should have "no open bugs" defined as its "go live" criteria. In other words, before your website is completely available to the public-at-large, fix all problems and errors found on your site. The best way to accomplish this task is to create and maintain a "bugbase" file (see Fig. 24).

Place all reported bugs in the bugbase as they are found. As each bug is fixed on the development server, launch the fixed elements and pages to the staging server as a new version of your site. Once that is completed, change the bugs' status in the bug-

Bugbase File

Bug Description	Date Fixed	Pending (y/n)	Date Closed	Deferred W/Explanation

Figure 24. Example of a bugbase file.

base file to "pending." Finally, verify the "fixed" bugs and either re-open them (if they weren't fixed) or close them (if the fix was verified). The final step is to document your testing procedure detailing the testing activity, remaining bugs, and their status (deferred for later fix, feature requests, etc.).

✪ SERVER ERROR LOG

One of your main aids in tracking down bugs is your server's error log — look for such items as missing images, bad links, and errors from CGI scripts, just to name a few. Use this excellent tool on a regular basis and after every update to your site.

Website errors mean lost traffic. It is normal to have a few errors in your log files, but if the error rate starts to rise precipitously, then look for what's causing the problems and fix them.

✪ OTHER ITEMS TO CHECK

Spell Check

Don't forget to run your spell check or manually to copyedit your website's text. Errors in grammar or punctuation jeopardize your credibility as a reputable e-business.

Hang Time

Measure the various browsers' "hang time." Note how long each browser configuration hangs with a blank screen before loading your web page. Is the hang time acceptable, with all configurations? If not, correct the situation.

Printing

Test how your website prints using different browsers. This is particularly important if your website is brochureware, subscription-based, or is a supplement to your technical help desk.

Some browsers can start a new page if a table won't fit on the current page; however, tables can force page breaks when the document prints out. By using tables you can control how the pages will print.

Avoid browser-specific code since your website may not print out properly if the customer uses a different browser.

Other printing problems can occur if you use images, or a black, or a colored background. Frames seem to constantly cause printing problems. If you think that your website's visitors will be printing your pages regularly, you might want to provide a non-frame version.

One common mistake is that contact information, support, and the Frequently

Asked Questions (FAQ) pages are not easy to find. Always keep those links clearly visible on every page.

Note: For those readers who are curious about the coding used to design a specific website you can do the following: If you have a Microsoft Internet Explorer or Netscape browser, just click on "View," which is located at the top of the browser, and then click "source" (page source for Netscape) in the drop-down menu.

✪ TEST AND RETEST

Okay, you've tested the design/front-end of your site and now you are ready to put the site on your staging server. Test it again. Make a hard and fast rule to always test every time you make any change to your site just to be sure you haven't accidentally introduced an error.

Other steps you should take during the testing stage include asking other people to test your website and to proofread the text. You and your staff are just too close to the project to notice what would otherwise be embarrassing errors.

If you developed a site that exists simultaneously in different formatted versions — flash/HTML, frame/non-frame — then, depending on the browser, you must test every version with every browser and with all of the possible option settings.

✪ CONTINUAL QUALITY ASSURANCE

Once your site is "live" and you have customers, don't drop your QA Plan, it is just as important now. With frenetic deadlines and constantly changing requirements, it is very easy to ignore the basics of quality assurance. An active, dynamic site means your web pages will be in a continual state of change. That means that, in all probability, the pages will be revised by different people; those people will bring their own individual coding quirks, which can and will introduce HTML and other coding errors (although I am sure all will try to follow the HTML on the templates).

If your website is currently operational but was brought to the Web without a QA Plan, gather your entire staff together (including consultants) and discuss ad hoc processes and standards that you currently have in use. In these discussions find out what causes the most problems on the site. This will give you the basis upon which to set up a QA Plan to address those issues as well as HTML and other coding guidelines. It generally is not practical to try to "update" your older content to adhere to the new QA Plan, but do initiate the QA Plan for all new content.

Listen to Your Customers

Listen when you are told that a customer has problems with your website. When someone takes the time to report a problem, pay attention. For every person who took the time to inform you about the problem, there are probably about a thousand others who didn't — remember competition is just a click away. Be diligent, pay attention and you will be one step ahead of most websites when it comes to neutralizing the little annoyances. After all, are you building your website to drive customer traffic to your competitors or to enhance your bottom line?

✪ QUALITY ASSURANCE PLAN GUIDELINES

Map your customers' experience. How? Fire up your browser and pretend you're a customer. Walk through the customer's complete shopping experience — run numerous tests from the moment your customer first enters your site through the last step of entering payment information and shipping preferences.

* Test the website's ordering process, from the user's point of entry through to the shipping preference.
* Test the credit card verification system, as well as the billing system that comes afterward.
* Stress test all of the servers with simulated loads.
* Check the website's impact on any back-end systems.
* Test not only the servers, but also the personnel and processes that support the systems.
* Test email response times.
* Test customer usability through focus groups or random-sample surveys.

Test the offline side. How is the payment information received, verified and processed, including the receipt of payment into your bank account?

Test your servers using simulated loads to see how they stand up under the stress of heavy traffic. In addition to the products listed earlier, look into:

RSW e-Load (www.rswsoftware.com), a component of Empirix's e-Test Suite, is the easy way to test the scalability of a website running on Windows NT. RSW offers a 30-day free trial. You can use e-Load early in your website's development process to validate the scalability of the overall architecture and to avoid costly rework later. It can also be used to test and to tune your completed website under load prior to deployment. e-Load utilizes the Visual Scripts to emulate thousands of "Virtual Users." You can change the number and type of users on-the-fly to try "what-if" scenarios as you vary the loading conditions or application settings. Also e-Load's integrated real-time graphics and reporting capabilities allow for easy interpretation of the testing results.

Facilita Software Development's Forecast (www.facilita.co.uk) is a non-intrusive web server load testing tool that simulates users on a system. It allows you to perform extensive and realistic load testing before a system goes live. It supports the UNIX family — Solaris, AIX, SCO, Digital UNIX, and Linux, along with Windows NT.

Software Research, Inc's Testworks/Web (www.testworks.com) allows you to ensure the quality and reliability of your web pages prior to publication. Components included in TestWorks/Web, such as the XVirtual tool, can simulate thousands of hits against your website by fabricating hundreds of synthetic interacting users, thereby helping you to assure 100% load accuracy. You can run the product on most UNIX and Windows systems.

Now, test your website's integration with your offline business systems such as your in-house database, accounting, inventory management system, etc. by simulating numerous transactions. In doing this you're not only checking out the hardware and software but you're also testing your personnel and the entire process that supports your website.

Last, but not least, check your customer service processes by testing your email response time, your web-based call center, etc.

Never launch an e-commerce site, or make a site change or enhancement without solid quality assurance. A month of intensive, structured testing is the minimum for a large-scale e-commerce site.

✪ FOCUS GROUPS AND SURVEYS

These are not in-house focus groups — don't use employees and friends — you need to find your intended audience and have that group participate in these tests. The only way to do this correctly is to employ market researchers with Internet testing experience or to use a service, like OTIVO (mentioned previously). Such services can be expensive, but they are necessary because this is where you can discover, prior to opening your site to the general public, if you have any features that your website's typical customers will find problematic or annoying.

Focus group testing includes asking these potential users about their expectations and usage, such as:

* How often do you shop online?
* What items do you most often buy online?
* After viewing the first page of this website, do you know what's in this site?
* Do you feel that the information provided on the home page is informative and useful?

Give the testing company your customer experience map so the focus group can set up various scenarios typifying an average customer. The tester then should give you a report comparing the different approaches the individual members of the focus group used to navigate your website. In this way you can determine what will work best for your customers.

The company that you choose to conduct these tests should be prepared to provide you with a complete report at the end of the testing period. This report should include:

* A short executive summary.
* Description of methodology.
* Key findings.
* Recommended actions.
* Summary of user comments.
* The full results (including any handwritten notes gathered during the testing, should be made available upon request).

There are two advantages to taking this approach. One is that once you have the results from the focus group testing you can intelligently change content organization and navigation structure as needed. The other is that the results will help you to understand your website from a customer's perspective, allowing you to refocus your site's future development on your customer's experience.

You could possibly cut the costs a bit by using a formal random-sample survey which means the marketing firm would not need to have access to a computer for each individual but would call individuals, ask them to use your site, and to then fill out a questionnaire.

✪ CONCLUSION

Only through a good QA Plan can you determine how and in what order each aspect of a website should be tested. A good QA Plan injects testing into every aspect of an e-commerce project. This allows you to detect defects early, close to the point of insertion, thus defects won't infect work later on.

Don't underestimate the value of a good quality assurance plan. And, *don't* skip this step. Not only must you execute a quality assurance test prior to launch, but also after you make any type of enhancement to your site.

Chapter 12

Website Maintenance and Management

If you want a place in the sun, you've got to expect a few blisters.

—Anonymous

Everyone pines for those uncomplicated days when maintaining a website was simply a matter of updating HTML files. Alas, that is no longer the case; today's e-commerce businesses are more robust, complex and demanding than ever before. As a result, more and more e-commerce operators are adopting an effective website maintenance and management system that syncs with their overall business scheme.

Incorporate in your website's blueprint a system to refine and to update not only your web pages to keep them fresh and new, but also your entire infrastructure to take advantage of new and evolving technology. In doing so, you ensure that your website will provide the two most important features of a web-based business — a compelling customer experience and minimal administration costs.

✪ ESTABLISHING A MAINTENANCE AND MANAGEMENT SYSTEM

How do you establish this type of system? Use the Internet — take advantage of its dynamic and interactive attributes. Your first step should be to put a customer feedback form in a prominent place on your website. Such forms are invaluable in getting your online customers involved in the process of analyzing your site. Many of your customers will be eager to participate.

Take the proffered information and use it to ensure that your site always achieves optimum performance from *your customers'* point of view. Don't be afraid to experiment; but constantly monitor the results. In this way you can make your changes, measure the results against other experiments, and settle on the best-performing choice.

Another tried-and-true approach for maintaining and managing a website is to tie it into your marketing strategies. This allows you to measure the results of your marketing campaigns — whether a specific campaign is on-track — bringing in customers and sales.

If you are a click-and-mortar, your website strategy must be integrated into your overall business strategy. This allows you to closely monitor whether your website is meeting your stated business objectives.

If shortfalls are found, bring out the blueprint and re-evaluate your web business plan. You must not only define the process but also institute a methodology that provides the means for a continuous evolution of your website's objectives. Remember, as the new century progresses, and technology races ahead, you will continuously redefine the role that your website plays in your overall business goals.

The establishment, management, and maintenance of a website requires a significant investment of time and resources on your part. But don't let these tasks overwhelm you. The best approach is to start with the basics:

* Design your site for ease of modification.
* Avoid sloppy formatting, numerous and gratuitous image maps, and superfluous links on every page.
* Know who will maintain and update your website and have more than one person that can easily step in and take control.

✪ AVOID SIMPLE ERRORS

When updating your website, check that errors aren't incorporated into your updates. If a mistake does get by you, fix it immediately upon discovery. As a website grows and changes, your site's fault-free design can become riddled with mistakes. For instance, you may find that gradually:

* The site's navigation scheme begins to breakdown.
* Broken links become the norm.
* Individual web pages have become too long. People don't like scrolling, two screens per page should be the maximum.
* Too many graphics have been attached to one page.
* You haven't kept up with mundane updates, e.g. the site's copyright information is

out-of-date (2001 instead of 2004); contact information is inaccurate — you especially want to have accurate sales contact information at all times. (Consider using general position email addresses names such as info@yourwebsite.com or sales@yourwebsite.com, rather than individual names).

✪ SOFTWARE

There are tools to create and to implement a set of uniform formats and styles so that you can instantly create a new web page by copying and modifying an existing page. For example, the following tools can help in keeping all coding uniform, thus eliminating problems caused by errant code. (The products price will depend upon your site's needs.)

✱ CyberTeams' Website Director (www.cyberteams.com) is a highly scalable web content management and development system that is priced within reach of just about any e-business's budget. Its WYSIWYG editor and page layout templates optimize website creation and maintenance by enabling content contributors and managers to develop sophisticated web pages without knowledge of HTML.

✱ Refresh Software SiteRefresh (www.refreshsoftware.com. If your website needs frequent content change and/or is a dynamic database-driven site, this vendor's relatively inexpensive hosted content management system may meet your needs.

✱ Topfloortechnologies' Insertrix content maintenance software (www.topfloor-software.com) enables a website built upon dynamic content to cost effectively change its content in minutes, without programming.

✱ Webceo (www.webceo.com) provides a unified workspace for nine programs that can help a website operator to attain real results in search engine marketing, to perform intelligent web traffic analysis, and to maintain your website easily. Webceo provides all the help you need to pinpoint broken links and other errors on your site; to edit your web pages in the WYSIWYG mode; to upload HTML and other files to your server, and to monitor your website. If you buy the entire nine-program package, Webceo also can provide visitor analysis, live traffic analysis, handle your search engine needs (optimize your site for high rankings in search engines, submit your URLs, and check your rankings with search engines).

✪ LINKS MANAGEMENT

Another major part of a comprehensive website maintenance and management system is to vigilantly identify and fix broken links. A broken link occurs when a page or location (whether on your website or some far flung web page) pointed to by another page has been moved, deleted, or renamed. Unfortunately, there are no automatic repair mechanisms standing by to sort the problem out. Most often the source page is simply left with a broken, or dangling link that leads to nowhere.

Any way you look at it, web-based businesses are heavily penalized for both broken outbound links (which are arguably under their direct control), and broken inbound links (to which a number of unsatisfactory solutions apply). Some obvious consequences of a failure to deal with broken links on your site and broken links that once pointed to your site are:

* Loss of revenue as potential customers attempt, and fail, to follow a broken link to your site.

* Brand damage since a broken link on a website is as bad, if not worse, than a misspelled word in a brochure. It takes diligent oversight to guarantee that a web page that revealed no flaws when placed on the Web, remains flaw free (including no broken links) as time goes on. A business with a website with broken links and broken links pointing to that business's website might lead a consumer to the conclusion that the error is caused in some way by poor standards or bad administration within your business.

* Loss of productivity for the people who maintain your site. Broken links are a time-consuming headache and large sites may find it costly to keep up-to-date with broken link repairs.

The author notes that today the average website has one page in every four containing a broken link. It has been written that more than 10% of the total links on the Internet are broken. Just think, maybe one page in every four that points to your website could be sending your customers on a one-way trip down the drain!

Do you have links to other sites? If so, you must check these links regularly and update or delete them when the links are inaccurate. Be diligent about keeping your links updated so you can identify broken links quickly and pinpoint the pages you need to fix to keep your site at peak performance all the time. You don't want your customers to see a "File not found" message when they are looking to buy. There are many applications and services available to help you with this, including:

ChangeAgent (www.xlanguage.com/products/cagent.htm). This link management tool discovers broken links, investigates their cause, and repairs them in a single, integrated environment. It also can remove orphan files and reorganize a site while maintaining link integrity. ChangeAgent works well with other applications allowing it to replace broken URLs while avoiding reformatting or restructuring your HTML code. Cost — $200.

CyberSpyder Link Test (http://www.cyberspyder.com/cslnkts1.html). This website management program not only verifies whether a website's links are broken, but also provides content analysis services. CyberSpyder is designed to work with websites of all sizes — from the very small e-commerce site, to the large corporate site with thousands of links. Costs — CyberSpyder is distributed as shareware, as such it may be used free for up to 60 days for evaluation purposes, after that there is a small registration fee of $35.

Link Checker Pro (www.linkcheckerpro.com). A nice, inexpensive, no frills tool to check internal and external links. Link Checker Pro is easy to use and has many customizable settings. It allows reports of link data to be saved in HTML format that may be easily viewed in any Internet browser. Costs — Free for up to 30 days, thereafter there is a $40 registration fee.

Link Sleuth (http://home.snafu.de/tilman/xenulink.html). This free spidering software checks websites for broken links. Link Sleuth's link verification is carried out not only on "normal" links, but also on images, frames, plug-ins, backgrounds, local image maps, style sheets, scripts and java applets. It displays a continuously updated list of URLs, which you can sort by different criteria. Costs — Free.

LinkScan (www.elsop.com/linkscan/). The LinkScan family of products provides a host of industrial-strength link checking and website management tools. For instance LinkScan offers flexible test automation capabilities for the entire spectrum of web-based applications. Costs — $750 to $7500.

Linkscan also offers a free service, called LinkScan/QuickCheck (www.elsop.com/linkscan/quickcheck.html), which provides online link checking and HTML validation service. This free service permits you to check up to 10 web pages per hour and up to 50 per day (although there is a limit of 200 links per document). QuickCheck is intended as a quick way to check the quality and identify problems on a web page.

If you are diligent, and use one of the above-mentioned (or similar) software packages with intelligence, your customers, when trying to access your site through a link or browsing your site, should seldom see some of the common error codes listed below:

Error 401 - Access to this page is denied.

Error 402 - A payment is required.

Error 403 - The request you have asked for is forbidden.

Error 404 - File not found.

Error 408 - Server timed out waiting for request.

Error 500 - Internal server error.

Error 502 - Error response received from gateway.

✪ CONTENT MANAGEMENT

Managing the content of a busy website is crucial. For example, size may be a problem — it is not unusual for a site to routinely publish hundreds of pages per week. However, the people creating all of this content are usually writers and marketing types who rarely consider the need to clean out dead files let alone understand the problem of tracing a chain reaction of errors caused by a simple change in code. Then you have those businesses whose entire website consists of constantly changing content — newspapers, ezines (an online magazine), B2B vendors, and catalog sites. These sites require a tight control of style and format, and easy re-purposing.

E-commerce businesses are encountering increasingly complex logistical problems coordinating their site's updates. Large and enterprise sites have web teams consisting of software and content developers, webmasters, graphic designers, testers, and approvers. Frequent changes by such a diverse team make it increasingly difficult for an e-commerce business to ensure high quality, especially since most changes are made at Web speed.

✪ HARDWARE AND INFRASTRUCTURE MAINTENANCE

Time-to-market pressures have forced many web-based businesses to create web infrastructures (everything that makes your pretty pages work) in a haphazard fashion. These websites sometimes experience substantial downtime as a result of poor planning and/or maintenance — a flaw immediately apparent to customers and business partners. Keeping your website up and available is the goal. Attaining 99.999% availability ("five nines") is an exhausting, arduous, and never-ending pursuit. Your website's failure points can include: human error, hacker infiltration, faulty software in routers and switches, increased bandwidth traffic that crashes servers, configuration problems, power failures, major carrier outages, not to mention any of the numerous applications that run your website.

The total "end-to-end" performance of the website's infrastructure must be understood and analyzed in order to ensure that it delivers the performance demanded by today's web customers. However, this is not achievable by the simple addition of a website management component to the mélange of existing systems, network, and application management solutions already peppered throughout the average enterprise website.

But the case is not hopeless. Although a majority of all websites are now hosted externally, it is still vital that a website operator understands the importance of real time management and capacity planning. Many websites are sophisticated systems that incorporate transaction processing and any problematic component between the customer and the website can effect performance and reliability. Help is on the way.

A formal management and maintenance system for a web-based business ran on in-house equipment is crucial. This scenario requires that a robust management system must be designed and implemented, i.e., the full "end-to-end" system — application, network, connectivity and systems — if the site is to remain reliable and maintain cutting edge technology.

To accomplish this, all of your servers including your web server should be as self-documenting as possible. All programs and other source code should be published, and you should have a web policy document with goals, practice, management, etc. that is frequently updated and in an easily accessible (but secure) location.

Here is the help (again, the cost depends upon your site's needs):

NetMechanic (www.netmechanic.com). This all-in-one website maintenance service performs whole site link checking, HTML validation, load time analysis, and server reliability testing.

TIDF Maintenance Repair Shop (www.tidf.com/theIDF/wmrs/services.cfm). This maintenance service focuses on website assurance (monitoring, diagnostics and testing, browser compatibility checks), preventative maintenance and enhancement services (recurring updates, database maintenance and upgrades, site efficiency improvements including website infrastructure and general enhancements and tweaks), and website repair and emergency services (emergency website re-deployment, emergency assistance, hosting issues, database fixes and code repair).

Or use an outsourcing service like Elance Online (www.elance.com). Elance provides access to global expertise that is cost-effective and of high quality. There are a number ways you can use Elance Online to gain the website maintenance you need: Use Elance to search for the expertise you need; post your needs on the Elance website and wait for proposals to come flowing in; or contact Elance's Project Services for project scoping, vendor selection, and/or project management assistance.

With the proper management and maintenance system in place, a web-based business can monitor and manage applications, data, systems, and networks in an integrated, pro-active manner using sophisticated software tools. In addition, the data gathered and reported can be used for other functions, such as security, capacity planning, etc. However, when it comes to five nines availability there's no single answer, just as there is no single point of failure. While a website should implement best practices to improve its chances, there are just too many factors that can bring down your website. The only sure way to minimize such incidents is strict attention to planning and preparation.

Note: Maintaining a website can be time consuming, so you might want to outsource your website's management and maintenance chores. While there are many qualified experts that can help you manage, maintain and even fix a problematic website, it's up to you to be specific as to what you expect from the outsourcer and to understand the outsourcer's contractually obligated duties. Read Chapter 13 before going down this path.

✪ CONCLUSION

A website must be kept updated or its moneymaking potential is wasted. In this chapter, we've discussed various software and online solutions to aid in your quest for the perfect, problem-free website.

Most web operators will find that they need to rely on a blend of technologies to provide the proper website maintenance and management. Thankfully, the functions offered by website management and maintenance tools often overlap. Thus, if you look carefully before taking the leap, you may find it possible to save time and money by purchasing a single suite of tools that can handle most, if not all, of the tasks necessary for a properly managed site.

Chapter 13

Consultants and Vendors

*A consultant is someone who takes your watch away
to tell you what time it is.*

Every web-based business risks getting mired in an outsourcing dilemma as they seek to obtain valuable expertise that is either not present internally or in short supply. There are literally hundreds of capable consultants, vendors, and matchmaker companies out there clamoring for your dollars. For the typical web-based business to choose among the myriad of options available to it, requires that you be able to identify which consultants and vendors have the capability and commitment to help the web-based business to reach its goals.

This chapter contains information about consultant contracts, costs, types of consulting services available, when to outsource and when to do the work in-house, etc. Surprisingly, nearly all entrepreneurs and most of the brick-and-mortar businesses do not have a formal process for selecting consultants or vendors and most selections seem as random as using the flip of a coin to make the ultimate decision. But with an effective selection process in place your new website will meet the needs of everyone, including your customers.

✪ SELECTING A CONSULTANT

At some point, every web-based business will need to retain a consultant. Consultants are the "hired guns" of the Web industry that eliminate an endless number of problems in a timely and cost effective manner due to the wealth of related experience that they bring to the table.

Your specific talent needs may be for a website architect, a web designer, a web developer, a marketing expert, a planning consultant, a security expert, or some other form of technical assistance. Formalizing an effective selection process is key to obtaining the best services for your needs.

As part of the process of sorting out your expectations and selecting your consultant, write a summary of your project and objectives. Include a concise description of your website (either as it exists, or as you anticipate it to be). This summary can be a handy reference tool when contacting prospective consultants.

To begin, assess the project's specific requirements, review your objectives to ensure they are well-defined, then determine what your staff can handle and what you need to outsource by asking yourself and your staff:

* How much of the work can be done in-house?
* How much of the work is beyond your staff's capacity and must be outsourced?
* What is the budget?
* Are there cost savings? If so, what is the expected payback period (i.e., ROI of hiring a consultant versus trying to do it in-house)?
* What is your business' commitment to the project? (In other words, after you spend time in the selection process — which in and of itself can be costly — will the project move forward to completion?)
* What is the timetable for completing the project?

The Consultant's Role

After the project's objectives have been determined and prioritized you then need to compare those objectives to the types of services consultants offer. Including:

* A specialized expertise.
* Ability to objectively assess a specific situation.
* A temporary supplement to your staff and knowledge base.
* Technical and economic analysis of alternatives.
* Development of recommendations.
* Design and programming support.
* Assistance with hardware/software selection.
* Assistance with implementing operational changes.
* Completion of one-time projects.

Consultants can help improve your website's operations and productivity, but don't forget that you and your staff are the experts in the operation of your web-based business; a consultant should enhance that expertise, not substitute for it. Do not depend on a consultant for decision-making — including purchasing decisions. Not uncommonly, consultants sometimes receive a commission when you (as their client) buy one of their recommended products or hire someone the consultant recommended.

The Consultant's Qualifications

After deciding to take advantage of a consultant's service, begin the process of identifying the type of consultant that will best meet the needs of your project. To find the right consultant:

* Re-visit consultants you have used in the past, assess that consultant's capabilities and limitations before deciding if you need to find another consultant.
* Obtain referrals from other similar web-based businesses, trade associations, or consultant referral services.
* Contact prospective consultants to identify their expertise, qualifications, and interest in your project. Request their marketing materials, including relevant educational background, experience, and professional certifications.
* Obtain references and a list of previous clients. Checking past work performance is one of the best ways to evaluate a consultant.

When checking a consultant's references and previous clients, find out whether the consultant has worked on projects similar in size and nature to the proposed project; has met the stated work and project deadlines; and he/she has been responsive, available and trustworthy.

On a more personal level ask:

* If there were any problems and, if so, were the issues satisfactorily resolved?
* If it was easy to work with the consultant?
* Whether the consultant was knowledgeable. What was the overall impression of the consultant?
* Find out if the final cost seemed in line with the original estimate. Ask if the final cost was more than the consultant's original quote. If so, how much and why?

The Request for Proposal (RFP)

Once you have narrowed your selection down to three to five perspective consultants, prepare a Request for Proposal (RFP). A RFP is a formal request to the consultant describing everything you want the consultant to accomplish and requesting the consultant to write a proposal outlining how he/she would go about meeting your demands and the costs thereof. The RFP can be as informal as an email or a telephone call, but it is best presented as a formal written document. The more information it conveys, and the more specific your requests, the more the RFP enhances the consultant's ability to draft a relevant proposal.

There are basically two types of RFPs. A defined, rigid proposal wherein is laid out, step-by-step, your goals and requirements and a more informed, creative proposal, a three to four page document simply setting out what the job will entail. The defined, rigid-style RFP allows for an easy comparison of costs, approaches and other criteria submitted by the consultant during the selection process. A creative proposal gives the consultant less structure thereby allowing for a greater diversity in response, but this approach can make the proposals more difficult to compare. Still, the creative proposal allows you to observe how the experience and knowledge of the consultant can be creatively applied to provide an exciting and unique approach to the project.

The Interview

After compiling a list of consultants whose skills fulfill your requirements, invite each consultant in for an interview. Be sure to:

* Explain your objectives and how you view the division of labor and responsibilities between the consultant and your staff. Be specific.
* Solicit the consultant's opinions — an outside perspective can be valuable.
* Set out your timeline.
* Ask the consultant to specify who would be working on the project.

If the meeting appears to be successful, take the next step — provide a Request for Proposal (RFP) as discussed above and obtain the information necessary for you to do your due diligence. Such as:

* Obtain a clients' list from the perspective consultant, including names and phone numbers.
* Find out how long the consultant has been in business.
* Ask what kind of projects the consultant has handled.
* Obtain samples of the consultant's work.
* Ascertain what resources the consultant has to complete the project, (i.e., staff, sub-contracting, etc.).
* Ask the consultant if there are any foreseeable problems in meeting your time schedule.
* Find out how the consultant charges — is it per hour, per day, or per project, and ask if a deposit is required.

In any communications thereafter, provide a written answer to additional information requests by the consultant — clear communication is important, and it ensures that you get the results you want.

Review the Proposals

Review the proposals received from each consultant and compare them to your established selection criteria to determine whether:

* The consultant responded to the principle needs based on the RFP's outlined objectives.
* The services set out are specific to your RFP and are clearly defined.
* The timetable covers both the consultant's time and your staff's time and that it is reasonable.
* All fees and costs are clearly defined, the billing procedures are specific, and the consultant's fees seem reasonable.
* The consultant has clearly defined the division of responsibility between your staff and the consultant's staff.
* The consultant sets forth the consulting personnel assigned to the project, including résumés, experience, and billing rates.

⊛ THE CONSULTANT CONTRACT

To protect your investment, get a written contract. Some of the specific issues that should be addressed in the consulting contract or letter of agreement are:

* Does the consultant use subcontractors? If so, does the consultant receive a commission for their services?
* Who provides the necessary insurance coverage? What type of coverage does the consultant have?
* How will unforeseen costs be handled? Is your approval required before such costs are incurred?
* Where will the majority of the work be performed? If at your place of business, is there adequate space?
* If information accessible to the consultant is confidential, a nondisclosure clause must be included.
* Include specific dates for the project's milestones, for each report that may need to be submitted, and for the project completion.
* What is the procedure for handling problems as they arise?
* How will revised work and costs be determined?

Work Plan

Most consultants bill on a time and material basis. Therefore, to protect the interest of both you and the consultant, set out a reasonable work plan with a not-to-exceed (NTE) cost. A work plan is like your blueprint, it maps out exactly what the consultant has contracted to perform, with timelines and estimated costs at defined stages. Use the proposal received from the consultant as a guide when drawing up your work plan. In some instances the work plan may vary due to the nature of the project, so a series of possible scenarios and costs should be built around best case, worst case, and most likely scenarios.

❂ WEBSITE DEVELOPERS

Many new websites are contracted out for development to website developers, an independent contractor performing a "work for hire" service. In this case you will look for a person or entity with both the technical and conceptual ability to be responsive to your requirements. A website developer will also need to produce a web design that is compatible with your business image. During the selection process, follow the procedure set forth in the "Selecting a Consultant" section.

You will also need to produce and sign a Development Agreement addressing issues that cover the development of your website including payment and acceptance procedures. The bare bones of a development agreement should contain:

* Firm dates for specific design milestones and website completion.
* A budget covering everything from the start through completion of the website including specific payment milestones.
* The developer's use of subcontractors and any commission the developer might receive for contracting with them for their services.
* The procedure for handling any problems that might occur.
* A mechanism allowing for revised work orders, change orders, and determination of the ensuing costs.
* The maximum acceptable download time for any web page.
* The inclusion of a user option, if necessary, of a low-graphics version of the website in order to minimize download times.
* A guarantee that the website is downwardly compatible with a specific version of Internet browser software such as Microsoft's Explorer, Netscape, and WebTV browsers.
* A guarantee that a specific number of users are able to simultaneously access the site, as well as setting forth a minimum response time.

＊ Assurance that the site will integrate properly with your business network, intranet, or other data server infrastructure.

＊ Assurance that additions, corrections, or modifications to the website can be made by you and your staff without interference with website operations.

＊ Specifications of the security safeguards, procedures, and firewalls that the site must contain.

＊ Guarantees of the functionality of online credit verification and acceptance procedures.

＊ Assurance of the scope and procedure for you easily to access, record, and compile information about the site's visitors and customers.

＊ A provision for timely documentation and source codes for all software associated with development of the website.

＊ Training of your staff to use and maintain the website's software including upgrades.

＊ Delineate the responsibility for transferring and installing the completed website to the web servers that will run the new website. This should include all software — either purchased or licensed by your business.

＊ Set out a provision for alternative interface designs that the website developer must provide for your review.

＊ A commitment by the developer to a set period for joint beta testing of the completed website and a subsequent specific number of days to be used for evaluation of the new website by you and your staff to ensure your new website performs in accordance with the agreement.

＊ The developer will promptly correct any bugs and failed links, including setting forth the maximum time for correction.

＊ The developer will promptly perform any revisions of the website that are necessary to comply with the functionality specifications.

＊ A right of rejection of the website if it does not meet the written specifications, setting forth certain options you may exercise regarding corrections at the time of a rejection.

＊ Any particular warranties or disclaimers by the developer.

＊ Assurances that any software for the site is free of any viruses or disabling devices.

＊ Conditions under which you have the right to terminate the developer's agreement and the liability of the developer upon such termination.

＊ A nondisclosure clause must be included, if information accessible to the developer is confidential.

＊ A copyright notice to be displayed on each page of the website.

The Contract

Once you have successfully negotiated the Development Agreement, to legally bind and protect all, a contract with the website developer should come next. It must address several issues:

*   Incorporated into the contract should be all of the provisions of the Development Agreement by reference.
*   Ownership of intellectual property relating to the content, screens, software, and information developed including who owns the rights to use any materials or software the developer creates for the website.
*   In development of your website, the developer must covenant that it will not infringe or violate the copyright and other intellectual property rights of any third party.
*   Normally you will receive a perpetual, irrevocable, worldwide royalty-free transferable license to any intellectual property, but typically the developer owns, uses, and retains ownership of the same.
*   Set out who is responsible for securing necessary rights, licenses, clearances, and other permissions related to any graphics or other copyrighted materials used or otherwise incorporated in the website.
*   The developer covenants that it will not ever use any trademarks, service marks or logos owned by you and/or your company, except with your express written approval.
*   The developer will comply with all applicable laws.
*   The developer will maintain satisfactory insurance and will provide proof of its policies.
*   A copyright notice will be displayed on designated parts of the website.

Other issues you may want to incorporate into the developer's contract or set forth in a separate agreement:

*   The developer will submit the appropriate information about your website to a specified list of search engines and directories.
*   The developer will not use its service affiliation with you for its own promotional purposes without prior written consent.

A website developer will be one of the first consultants you will need for your new website. Go slowly, follow the steps outlined herein, choose carefully, and you will be on the road to a great website.

✪ VENDOR SELECTION

A vendor has a product for sale that may require customization for you to utilize it. Vendors sell you the product and their professional technical services, as a package. It is imperative that the vendor selection process not be handicapped by internal political agendas, "gut feelings," or the toss of a coin decision. Once it is known that you are "in the market" you will be deluged with specific vendor related solicitations. If not managed correctly, the vendor selection process can eat up time and resources and it can be quite costly, accounting for as much as 25-30% of the total vendor costs.

Some of the worst-case scenarios can be:

* Choosing a vendor without proper due diligence. For instance, the vendor might go out of business in the middle of your project resulting in you being forced to make an ill-timed decision in order to complete the project.
* Service and support problems.
* Differences in your visions that were not made clear.

Vendor Evaluation

Establish a vendor evaluation methodology that does not focus most of the attention on the functionality and cost issues. A vendor's ability to execute a specific plan and its long-term vision is well worth the investment. The next step is to create a vendor's RFP.

Elements of a vendor RFP:

Functionality: This is usually the primary focus of any vendor evaluation, but in actuality it should represent no more than 40% of the total decision-making process.

Storyboard: To ensure that you cover all of the capabilities you want from a specific vendor product, use a storyboard. Draw the storyboard using the same criteria as set out in Chapter 2. In other words, include:

* The design aspects.
* Ease of use features.
* Ability to integrate with other tools, both hardware and software.

You must also determine if the product provides:

* Tools that enable you to present content electronically.
* Configurators to easily target various vertical, and geographic markets.
* Scalability to incorporate new media formats.
* Ability to easily interface with other e-commerce applications.
* Ability to enhance rather than decrease the security risks.

Costs: Don't put too much emphasis on the initial cost of the product in your decision-making process. Remember that the majority of the cost is hidden — such as product training, customization, and integration. Therefore, cost calculations should always include not only initial license costs for the product and any knowledge tools utilized, but also costs of installation and maintenance, help-desk gateway, ongoing education and training, and professional services such as customization and integration.

The Vision: What is the vendor's stated and realized development plans for its product? Ask how the vendor plans to incorporate and utilize new technologies into its product's architecture and how it plans to evolve its current product by adding to or enhancing the current functionality. Let's not forget about service and support — how does the vendor plan to grow and change its general and professional services support. Remember to investigate (due diligence once again raises its ugly head) how the vendor treats its customer once the vendor has been paid.

A vendor's vision is just as important as the vendor's ability to produce a viable product and execute its successful implementation. For most businesses, this can be used as the key differentiator when making the final vendor selection.

Service and Support: A product's functionality and cost provides, in many instances, the inducement for making a specific vendor selection. But it is just as important that you also consider the availability of quality service and support, for without them, the success of any product implementation is ultimately doomed.

Service and support needs to be broken down into two areas. The first is general support including installation of the product and continuing support. Also don't forget to check out the quality of the vendor's help desk services. The second area is professional services. This can be the weakest link in many vendor organizations and therefore due diligence must be stringent — i.e., evaluate the vendor's strengths in project management, systems integration, and business consulting skills.

In spite of the critical service and support that vendor's supply, most businesses don't take the time to create a vendor RFP — let your business be the exception.

Due Diligence

While we have harped on performing proper due diligence throughout this section, it has proven to be the weakest area in most vendor selection processes. Given the maturity of the e-commerce arena and the fierce competition that exists, the author predicts that, in a few years, at least half of the current vendors will no longer exist. Have you taken this fact into consideration when making your vendor selection? If not, do so. What is the vendor's financial viability? Answer this question by analyzing the vendor's revenues, growth, margins, sales and marketing investment, quick ratio, etc. The next step is to measure the quality of personnel within the Vendor's sales and development departments. Ask questions, such as if these departments are able to meet industry milestones; have historically been able to meet deadlines and delivered what they promised; have R&D capabilities comparable with other vendors; and is the vendor suffering a high turnover among its most talented people.

Finalizing the Selection Process

Once you have put in place your selection criteria and structure, use your storyboard to guide you through each step in the selection process. This will help to point out your website's needs. But you must also research, so as to pinpoint the exact requirements you expect a vendor's products to provide. Then, write the RFP, and send it out to a select group of vendors. That's straight-forward, right?

The next step is to whittle down your vendor selection list. This can get a bit hairy — use RFP validation, comparisons, scripts (the storyboard) and vendor presentations to help in this process. In the end, hold a no-holds barred in-house meeting and review everything, you should wind up with a short list of no more than three vendors that fits all the criteria.

You can now take the final step — negotiation and selection. Here you develop and sharpen your negotiating strategy. How? Decide what's critical to the project and put it all down in a contract (use your legal team for this part). Call in the short list vendors and the race is on. Once you have selected a vendor willing to work within the negotiated terms and conditions, you can begin the final documentation, presentation to management, and getting those signatures on the dotted line.

Following the advice in this section should usher your web-based business into many successful vendor partnerships.

✪ CONCLUSION

By following the procedures outlined in this chapter, any e-business should be able to utilize the best evaluation criteria, gather the necessary and objective data, and guarantee that its overall evaluation process proceeds in a structured format. You will also end up with a great paper trail, which can be useful if there is a need to explain how you arrived at a specific selection.

Chapter 14

Web-hosting Services

Technology. . .is a queer thing; it brings great gifts with one hand,
and it stabs you in the back with the other.

—C.P.Snow

There are a multitude of web-hosting services (a third party that sells you space on its web servers) available. A majority of small e-commerce businesses will contract with a web-hosting service rather than use an in-house server.

Your web-hosting decision is one of the most important decisions you make when starting an e-commerce business. That is because a good web-hosting service provides a place where your site can reside and operate quietly and efficiently in the background. But, if you don't do your research and perform proper due diligence, you might choose the wrong hosting service. The trick is to find the right match between what your e-commerce site needs and what the web-hosting service offers.

In your quest for the right hosting service, start with the basics — how good, how reliable, and how industrial-strength is your potential host's connection to the Internet? Then check out what other services are offered to enable your e-commerce business to have more functionality.

As you shall learn, there are a multitude of web-hosting business models and options available to an e-commerce business. Because there are so many hosting choices, you must decide what your own web-hosting requirements are before venturing forth. This allows you to narrow your shopping down to just a few hundred rather than thousands of choices. Let's first look at the various web-hosting models. Understanding the different business models under which web-hosting services operate makes it easier to choose the best service for your e-commerce site.

◯ THE WEB-HOSTING SERVICE MODELS

Web-hosting services are *not* equal. Although, at one level, all hosting services provide hard disk space on powerful computers (the servers) that have 24-hour connections to the Internet, there are many differences. By breaking this business group into different business models, the differences become clear.

Internet Service Providers (ISPs)

Throughout the world, thousands of local and national ISPs provide dedicated permanent connections to the Internet backbone and allow individual users access to the Internet (for a fee) via their connections. Some ISPs also provide some type of ad hoc website hosting service. Many ISPs, however, don't really understand the needs of a small web-based business. Nor are they quick to offer improved service because their bread-and-butter is dial-up access and that's where they concentrate their focus and investment. Furthermore, although a few large ISPs have a separate division that specializes in website hosting, many ISPs just don't have an adequate infrastructure for web-hosting. If possible, pass on ISPs and look elsewhere for your web-hosting services.

Sub-domain or Non-virtual Account

What 99% of you don't need or want is the 4 or 5 MB of web space that you can get for a small monthly fee or even free from an ISP when you sign up for a regular Internet dial-up account. This type of service is referred to as a "sub-domain" or "non-virtual account." It allows you to build a very small, simple website that can be accessed by a URL that looks something like www.mystupidsite.ISP.com or www.ISP.com/~mystupidsite/home page.html which means you have a directory on your ISP's website in which you can build your site. The profound drawback to the option is not only do you have a hard-to-remember URL for your website, but if you ever move the site, you'll have to change the site's address. This type of web space is not something to use for a serious commercial website.

Dedicated Web-hosting Service

A dedicated web-hosting service provides you with server space (virtual, dedicated or co-location) for your website, although usually without Internet access. Their primarily business is offering server and rack space for businesses that do not want to manage their web-based servers in-house. While a few web-hosting services also might offer

ISP service as a convenience to its core group of customers (e-commerce businesses); they probably are not going after the "casual surfer's" business.

Thus, one advantage of a dedicated service is that your customers won't be sharing valuable bandwidth (a big concern when choosing any type of hosting service) with the casual web surfer checking their email, downloading MP3 files, etc. What this means is that your customers will experience speedy access times, fast loading pages, and no shopping cart lag.

All large or enterprise websites should consider this type of web-hosting service. E-commerce sites that plan to incorporate streaming media or other types of services that require specialized servers should also turn to dedicated web-hosting services.

Since this market is over-supplied, to stand out in the crowd, some web-hosting services offer very nice (but expensive) high-end features. You will find that a good full-featured web-hosting service has a more "services" oriented approach — offering expertise in networking, or offering and managing complex software — as opposed to just selling space on its servers' hard drives.

Local Web-hosting Services

Check out the web-hosting services in your local community. These services can offer personalized attention and customized service, such as visiting your business in order to evaluate your hardware, examine your network, etc., and then making specific recommendations. For a small web-based business, a local web-hosting service will, in many instances, provide added services including, for example, helping you market your business by providing a link to your website on its local business directory. Therefore, many small- and mid-sized e-commerce operators may be willing to trade off certain features offered by a larger hosting service for the extendibility of a local service. Note though that a local web-hosting service is less likely to provide 24x7 technical support.

If you choose one of the small local hosting services, and your site becomes very active, don't be surprised if the host asks you to move your website to another service. Why? Because a good hosting service, even if small, will look out for the welfare of its overall constituency and might be afraid that heavy traffic to a particularly popular website on its service will hurt the performance of the other websites that it hosts.

Web Developer Hosting

Using your website developer to host your e-commerce needs is another viable option for the small- to mid-sized e-business. Quite a few web developers have added web-hosting to their list of client services.

A web developer usually provides hosting services via a server with a small pipeline to the Internet (either DSL or some type of partial T-1/E-1 service). Many times an e-commerce business, especially one with a small niche market, will find that the great service provided by such a hosting arrangement is more important than a large pipe to the Internet.

A web developer's business is, by nature, customer-centric and thus it will usually have the staff to help you with problems that might arise. However, there is a price to be paid: a slightly higher contract price, total dependence upon one service, and lack of 24x7 technical support.

One piece of advice: If you decide to go this route, be careful to ensure that your hosting contract does not lock you into using only the web developer's services (hosting or otherwise) for an extended period of time. Also, take the appropriate steps to ensure that the developer's name isn't listed as the "Administrative Contact" for your domain name. This identity confusion can cause problems and delays if you (at any time in the future) want to transfer your website and domain to another hosting service. You can find out who is currently listed as Administrative Contact for your domain name at www.accesswhois.com.

Free Web-hosting Services

Although free web-hosting services aren't as prevalent as in the past, you still want to avoid placing your web-based business in their hands. Here's why:

* You will be severely restricted in the things that you can do with your website.
* You won't be able to use your own domain name. You'll have to settle for something like www.freewebhosting.com/yourcompany.
* Although free web-hosting services normally give you between 2 and 10 megabytes of server space, many require you to display their advertising banners (or put up with annoying pop-up windows carrying their ads) and many won't allow you to display advertising from any other company.
* Most won't give you the easy-to-use FTP services to upload your web pages. This means you will have to find another way to put your changes on the web server.
* Few, if any, of them provide database access.
* Very few will give you CGI-bins.
* Few, if any, provide secure servers for credit card purchases.
* Forget about data backups.
* Search engines will normally take three to four times longer to list the submitted pages. Also search engines give more "weightage" to sites that have their own domain names.

* You won't get unlimited autoresponders, nor the POP email accounts you need.
* Most free web-hosting services will not give you access to your own server log files.
* You and your customers will experience frequent downtimes because of the heavy loads on the hosting company's servers.
* You will find that technical support is almost non-existent.

Get the message? Don't use a free web-hosting service for your web-based business — EVER.

✪ TYPES OF HOSTING ACCOUNTS

As you have noticed as you've read through this chapter, when it comes to web-hosting services, one size does not fit all. The same goes for the type of account you opt for with one of these services.

Virtual Hosting

Most small e-commerce websites should consider "virtual serving." This type of hosting service lets you run your website as if you had your own in-house web server but with the advantage of the web-hosting service's pipeline to the Internet. Virtual serving just divides a server's capacity into several distinct "virtual" web servers. With this set up a hosting service can host several sites on one computer. This type of service also allows you to load your own software, set up your own cgi-bin directories, etc.

Most Internet hosting companies offer several varieties of "virtual serving." If you require a more powerful server or a server with a limited number of sites sharing it or a combination of both, you can get it (for an appropriate fee). You are only limited by what hardware configuration and bandwidth you can afford.

Your Own Dedicated Server

Another option for a small e-commerce site that has heavy traffic, needs high availability, or serves dynamically generated pages is to contract for a dedicated server. With a dedicated server contract, instead of sharing a server, you run your website on one of the web-host's servers, which is configured for, and dedicated to, your website's needs. This lets you take advantage of the web-hosting service's high-speed connection to the Internet, technical support, and redundancy systems, but allows you to have, at your command, the full capacity of an individual server.

A dedicated server is the right option if a virtual hosting service plan doesn't offer enough data transfer, disk space, CPU power, and/or flexibility. With a dedicated server your website has the whole server to itself. That means high data transfer limits, hard drives with gigabytes of free space, 100% of the CPU power, root access, and the ability to run whatever programs you wish.

Dedicated servers usually come preloaded with an operating system, web server software, control panel, and some basic services. Anything else to be loaded onto the server is up to you. Most dedicated servers are rented by the month and pricing usually includes a monthly bandwidth limit and at least one IP static address.

The downside of this choice is that the cost of running your website on a dedicated hosted server can be significantly higher than using a virtual server. Also, you need to know how to manage a server. That includes monitoring, installing and upgrading programs, configuring programs, dealing with hack attempts, troubleshooting and fixing problems, etc. Or you could hire an experienced system administrator to handle such tasks.

Another option is to let the web-hosting service provide your server management needs. For an extra fee you can get a *managed* dedicated server — many Internet hosting companies offer leases for individual computers with the same management options as a virtual hosting contract.

Co-hosting or Co-locating

A large website might go with co-hosting (which is also referred to as co-locating) since it provides the most freedom — but it is expensive. With co-hosting you rent space in an Internet hosting service's network server cabinet and pay to access the network that's connected to the Internet. In this situation you put your own computers in the server cabinet and service them yourself by obtaining access to the hosting facilities. This arrangement usually includes, for a fee, some kind of limited maintenance and back-up service.

Basic Server Farm

A basic server farm usually consists of several separate servers performing different functions — all residing in one server cabinet. A large website running various applications should separate the various server applications over different computers, some sharing tasks and others going it alone. It's good practice, for example, to put a processor-intensive application on a single server and to isolate any risky tasks. See Chapter 4 for discussion of hardware configurations.

✪ DIFFERENT HOSTING NEEDS

There is a big difference between the web requirements of a small- to mid-sized web-based business' hosting needs (even if it is a full-scale e-commerce site) and an established large- or enterprise-sized business moving to the Web with a huge catalog of products and heavy traffic.

Whichever category you fall into, shop around for your web-hosting service provider and thoroughly check references. Learn to differentiate between a pitch and a promise. The first step in this process is to ask yourself what do *you* want from a web-hosting service. Then determine your website's needs.

The Small- to Mid-size Website

It is important that strict attention be paid to how your hosting service and your web-based business will work together, including how your chosen hosting service can compliment your web-based business.

Shopping for the right web-hosting service for your web-based business is a difficult task at best. Look for a web-hosting service that offers a basic hosting package that includes:

* Unlimited data transfer.
* At least 15 MB of space (if email, log files and system programs are included in the amount of MB offered).
* The option of FrontPage extensions, since so many websites are designed with FrontPage.

Numerous hosting services offer small business-specific hosting services. To help you in the task, make a list similar to the one depicted in Fig. 25. Use the list to start your comparisons of web-hosts.

Switching hosting services after your website is up and running is a headache. You can often avoid this pain by selecting, from the get-go, a hosting service that provides a full range of hosting packages — some of which you need now, and others that you will need in the future.

You will find in almost every hosting package a set number of email mailboxes, a predefined amount of server space for your files, and a traffic allowance (the number of megabytes downloaded when customers request one of your web pages). As with everything else, the more email boxes, file space, etc. you want, the more you pay. Unless your website is very well established and image-driven with, for example, a large product catalog, you can start with the one of the less expensive plans offered by your chosen web-hosting services.

Many web-hosting services also offer web design packages; but be advised that most of these packages leave a lot to be desired. If you do decide to use your hosting services' web design services, subject their in-house design team to the same scrutiny as you

Your Website's Hosting Needs

Needs	Current	Projected Needs at End of the 1st Year	2nd Year
Reliability	90%	100%	100%
Services you want			
- server space	5MB	20MB	25 MB
average small site needs 5MB)			
- # of email boxes	5	15	20
- email forwarding	no	maybe	maybe
- traffic allowance	??	??	??
- ready-to-use CGI scripts	yes	maybe	maybe
- tracking tools	yes	no	no
- tools to register with search engines	yes	yes	yes
- scheduled backups	yes	yes	yes
Cost you can bear	$3600	$5000	$7500

Figure 25. A list can help you to determine what you want from a web-hosting service. The reason there are three columns is that what you want and need now won't be what you need a year or two down the road.

would a web design firm or a consultant. If you design your website yourself using your hosting service's web design solutions, scrutinize the proffered web design tools carefully before moving forward. Most of the time they are inadequate. Your best bet is to buy and to use your own design software.

For most small- to mid-size web-based businesses, contracting with a full-service web-hosting service makes the best sense. Paying a relatively small monthly fee to a service to host your website on its servers, which are maintained by its technical people, and connected to the Internet via its pipe, allows you to ameliorate hardware costs and avoid hiring expensive technical staff.

Top web-hosting services that a small- to mid-sized e-commerce site might consider include Globat.com, ICDSoft.com, iPowerWeb.com, Jumpline.com, Lunarpages.com, SpeedFox.com, and WebsiteSource.com,

The Large to Enterprise Websites

For some e-commerce businesses, a large, national web-hosting service is the only

IN-HOUSE HOSTING

We have discussed the possibility of hosting your own small website in Chapter 4. Don't try this with a large website. It would be a nightmare unless your company already operates servers on a 24x7 basis with a 24x7 technical staff on the premises. The cost of the large bandwidth connection (pipeline) to the Internet is daunting. So, please find a place to put your web servers that has engineers on site 24x7 to monitor them. This will provide a much less costly and less problematic website.

It is difficult for many of you to forego the opportunity for hands-on experience with your servers, but once everything is correctly configured and running, there's very little to do. What you might need to do can be managed remotely whether you are running UNIX or NT systems.

choice. These e-commerce sites typically have very high traffic, provide server intensive features such as chat rooms and large media content (e.g. graphic files in their product catalog, streaming media, etc.). These websites not only has a need for, but have the budget to pay for, a web-hosting service that offers, at a minimum:

* OC3 or better connection to the Internet.
* Mirror sites.
* 24-hour staffing.
* Redundancy (including your connection to the Internet backbone).
* Robust security.

Expect to pay a substantial sum for 24-hour full service technical support since people cost more than computers. Yet, for most large websites, all high-traffic websites, and all enterprise sites, this type of service is mandatory. However, before you sign on the dotted line, be sure to checkout how much service you actually receive from the service's technical support. Determine:

* How many hours a day is technical support staff on the premises (it should be 24x7).
* How fast they respond to a request.
* Is there a response time guarantee?
* How much help will the technical staff actually provide under various contract terms?

An enterprise-class hosting service should provide a project leader and an enterprise management team to run your website's entire server infrastructure, including the firewall, on a 24x7 basis. This team should manage, support, and upgrade web applications, whether standard or proprietary, and add networking options, including a Virtual Private Network (VPN), Internet, frame relay, and dial services.

While the individual e-commerce client's enterprise-class hosting services will differ depending on the scope of the e-business, there are some features all have in common (in addition to what has previously been discussed), including:

* Dedicated, point-to-point, unshared T-1/E-1 or T-3/E-1 access to the Internet.
* Complete DNS services.
* End-to-end implementation management.
* Online network utilization statistics.
* Choice of customer premise equipment (CPE) solutions.
* A fully managed service.
* Redundancy, e.g. a T-3 or better connection to diverse Internet backbones, UPS and generator power backup, a redundant network operations center.
* High availability cluster multiprocessing consisting of a primary machine and a live standby. Data contained on the primary server is seamlessly mirrored on the standby machine. If the primary should fail, the standby server will immediately assume the role of the primary machine.
* Dedicated e-commerce that offers customized systems for purchases, services, and delivering order status and web application hosting.
* Web applications such as email, human resources, finance, and other operations to secure off-site servers with high-speed access and a single point of contact.
* Virtual Private Networking (VPN) and Virtual Private Dial-up Networking, enabling your company to connect its LANs, hosting sites, business partners, branch offices, telecommuters, and mobile employees in an integrated, secure, and simple manner.
* Managed software services enabling the enterprise IT organization to provide asset management, software distribution and management, network management, network performance tuning, and desktop management.
* Streaming media delivery through scalable, on-demand capability that distributes images to single and multiple locations to support videoconferencing, PC-to-PC conferencing, and video distribution services.

But that's just the beginning. The right enterprise-class hosting service will have trained system administrators, along with other qualified personnel, on site 24x7, monitoring your web servers along with the other web servers they host. The hosting service's technical personnel monitoring your site provide instant attention and follow pre-arranged instructions for any situation that arises. These instructions can be as varied as a call to a named technical person in your business, to handling the problem himself or herself, or anything in between.

It is well known in the industry that security is an entity that has a life of its own; therefore, leaving the matter to the security experts at a good hosting service is a wise choice. A top-notch hosting service is positioned to provide first class security through multi-staged access control and trained security personnel monitoring the equipment and the system, 24x7. Such hosting services also will provide a configured router and will offer front-line firewall solutions which you are advised to take advantage of since, they are neither easy nor cheap to set up.

Don't forget the mundane. An enterprise level hosting service should provide automatic backups of your data, and storage for the backups at an off-site storage location. Visit the physical plant. Is there sufficient air conditioning to keep the equipment cool? Is it clean and free of dust? Is it secure?

An enterprise-class service contract costs a pretty penny, but for some e-commerce sites, it is worth it. After all, what will it cost if your site goes down? The odds are that your website will suffer outages due to hardware failure, software glitches, outside attacks, and more. If you are self-hosting your site, this can be a huge problem. Do you know how you'll handle such situations? What will the cost be to your business when your site is down? Can you afford it?

The high-end web-hosting business is very competitive. These hosting companies offer sophisticated, large-scale, end-to-end solutions tailored not only to the enterprise customer's immediate needs, but engineered to respond to future demand, with an emphasis on customer support and service. If your e-commerce business is in need of this type of hosting service, check out companies such as Above.net, AppliedTheory, Digex, WebIntellects, Rackspace, and Verio.com to name a few.

✪ FINDING THE RIGHT WEB-HOSTING SERVICE

Whether you operate a small, large, or enterprise e-commerce business, shopping for a perfect fit for your specific e-commerce needs can be complicated. Start by browsing the web-hosting service's website.

* Is there a phone number, a real mailing address (not a P.O. Box)? If the service does not have a telephone number posted on their site as well as their hours of operation, ask yourself "why"? Maybe they do not want their customers to contact them. Reputable hosts will have their telephone number and other contact information posted on their site in a very visible location.

* Is their FAQ (frequently asked questions) section comprehensive, i.e. does it answer most of *your* questions?

* Does the site host an online users' forum for discussing the host's services? If so, read

a sampling. This allows you to gauge what problems other websites are having and how responsive the host was perceived to be to those problems. Then go to a search engine and look for any other information (e.g. reviews) that might be available.

With a little more research you can whittle down your list of viable web-hosting services even further. While costs are important, don't let it be the only deciding factor — many low-cost web-hosting services are simply not worth the money. *You get what you pay for!*

Look for a hosting provider that has quality equipment and offers a variety of features as well as the technical expertise to support their customer base. Consider choosing a web-hosting provider that offers features that you plan to add to your e-commerce site in the near future. Switching hosting providers takes a reasonable amount of effort and is not something you want to do simply for the sake of saving a few hundred dollars.

Find a company that has a few years of web-hosting experience. Ask when the business began its *web-hosting* operations, not its web design business, nor its ISP services Avoid web-hosting services that:

* Have daily data transfer limits (transfer rate fees should be based on monthly usage not daily).
* Hide the fact that they charge a setup fee (virtually all hosting services charge a fee).
* Offer PayPal only accounts (some, but not many, accept payment by PayPal only; credit cards are non gratis).
* Limit file size (not the total disk space but maximum size of a single file) to discourage downloads and large graphics.
* Flat out don't allow downloads.
* Have an FTP transfer limit.
* Don't provide an emergency telephone number.
* Place severe limits on resource use. While most hosts impose CPU limits on shared plans, be sure the proposed host doesn't forbid resource-intensive scripts such as message boards.
* Provide no technical support on weekends.
* Offer no control panel (especially if you are new to the Web).
* Place a limit on monthly site hits (some hosting services do this in addition to data transfer limits).
* Require that you sign a long-term commitment.

✳ Offer no money—back guarantee.

Now it is time to create a list of *your* website's specific requirements. Be sure to consider each of the following topics.

Windows or Unix?

Most, but not all, web-hosting services support only one operating system family either Microsoft Windows or Unix (which includes Linux). If you use or are familiar with Microsoft's IIS web server, Active Server Page, VBScript, SQL server or Visual Inter-Dev then you probably want to go with a Windows shop. If you lean toward Linux, FreeBSD-based operating systems, Apache web servers, or already have a good understanding of Unix-based solutions, then find a web-hosting service that uses Unix.

Note: Your business's internal operating systems don't need to match your web host's systems. If you run Windows XP and want your website to run on Apache or if you find a good web-hosting service that operates Linux, there will be no compatibility problems.

Size of Pipeline

The pipeline connects your server to the Internet — the bigger the better. But the more bandwidth provided, the more it costs. It is the mind-boggling expense of installing an adequate pipeline to the Internet that prohibits most web-based businesses from setting up their web servers in-house. High-speed access is obviously the key to a responsive website. If you expect your website to experience heavy traffic, find a provider that offers a minimum of T-3/E-3 (45 Mbps) connectivity and verify that there is sufficient bandwidth available for each client. This bandwidth should preferably be connected directly to the Internet's backbone (the worldwide structure of cables, routers and gateways that form the Internet). The more direct access your web server has to the backbone the less likely your website will suffer from a "data traffic jam."

The way these services charge for bandwidth varies. Most web-hosting services will charge you a monthly fee based on the amount of traffic your site receives. But some will connect you as directly to the backbone as possible and provide you with all the bandwidth you can use for one set fee.

While some small local web-hosting services (including web developers that provide hosting services) might offer just a DSL or "fractional T-1/E-1 connection to the Internet, most web-hosting services offer much more. The typical hosting service's computers are connected to the Internet backbone via a T-1/E-1 (at a minimum), T-3/E-3 (the majority of local services), or some type of OC line (mainly those catering to the enterprise crowd). For a complete discussion on connectivity, see Chapter 6.

All things being equal, your best bet is to look for a hosting service that offers at least a T-3/E-3 connection (unless your website serves a small niche market with limited traffic).

Note: One common problem is data traffic jams. Ask a potential web-hosting service what options are available to you if you find your website's bandwidth (speed of connection to the Internet) is compromised by another website's high volume of traffic.

Know Your Traffic Limits

Find out how many page hits (the number of times someone transfers one of your web pages to his/her browser) are included in the basic price quoted. This will probably be stated in terms of megabytes of file transfer (amount of data transferred out of your site); if so, ask the host to translate it into an average hit rate. Although some web-hosting services have no limits, others will apply a surcharge if you go beyond a pre-determined limit. Some hosting companies charge for the number of hits (the number of times someone transfers one of your pages to his browser). Others charge according to the amount of data transferred out of your website. Either way, the busier your site, the more you'll be charged under these kind of pricing schedules.

Note: You can estimate your data transfer needs by looking at your average web page size (including graphics) and multiplying that number by the number of page views you expect to have in a month. For example, if the average web page size is 50 KB, and you expect 2000 page views per day, your average data transfer rate will be about 3 GB per month. But in such a case, you should get a plan with a 4 to 5 GB of monthly data transfer limit to avoid overstepping your account's limit in case your calculations are wrong or your website's traffic increases. Moreover, if your calculations indicate that your site's average data transfer rate will be over 50 GB per month, look into dedicated servers rather than virtual servers.

Unlimited use may not be the nirvana you think it is. Furthermore, unlimited or unmetered accounts are usually just a gimmick — a web-hosting service would quickly go out of business if they didn't charge something for data transfer because they have to pay from $1 to $5 per GB of data transfer. However, that's not to say that a good web-hosting service won't provide a certain minimum data transfer for free. Most do and for many smaller e-commerce businesses, that figure is more than adequate. As a rule of thumb, a small e-commerce site will need, on average, less than 100 MBs per month.

Compare traffic allowances. Hosting services that limit the amount of material that can be downloaded from your site each month may hit you with a large surcharge when that limit is exceeded. If your site offers products/services that require download (software, music, white papers, tech support information), you need to find a service that doesn't set limits. Some offer huge traffic allowances (several thousand MBs per month) for the same price as others that limit you to a few hundred MBs per month.

Finally, be advised that if you sign with a web-hosting service that offers an "unlimited" plan, read their "acceptable use policy" very carefully. You will find that most state that if you use too much disk space, bandwidth, or CPU time, you must upgrade your service or leave. And remember to plan ahead — allow for future growth of your site and its traffic.

Space

Web-hosting services usually assign a client a defined amount of disk space for "virtual" service. In other words, when you're using a web-hosting service, most of you are buying space on their server's hard drives. You don't want to handicap your site with too little disk space, but at the same time you don't want to pay for unused disk space.

As little as 5 MBs may be plenty of space for the web pages and graphics of a brochureware site, as long as email, log files, and system programs are not counted in the 5 MBs. Otherwise, more disk space is needed because those items take up considerable space (as much as 15 MBs more).

As a rule of thumb, 10 MBs of disk space equals about 100 web pages, which is more than enough for the average small business site. Of course, if your website offers lots of images, sounds, animation, or applets, you'll need more space. Again, plan ahead — it can save you money later on. If you expect to grow, look for a place that offers sites of 100 MBs or more. Ironically, some web-hosting services offer 300 MBs sites for less money than others offering only 30 MBs.

Ask the hosting service how many websites are hosted on each of its computers. Although most hosting services are not too forthcoming about this information, you should be able to at least learn if the host has any policy limits.

Uptime

Check the hosting service's reliability or "uptime" — the percentage of time the web-hosting service is up and running. Most claim 99.9 percent uptime, but ask around. XO (previously known as Concentric), claims to have never been down. Digex also has a top-notch uptime rating.

Multimedia

If your website utilizes multimedia, look for a web-hosting service that supports the technology used by your website. This is important. You want your customers to be able to listen to audio or to watch video without the necessity of fully downloading the file.

CGI Scripts

Before signing with a web-hosting service you must know what server-side languages (also known as CGI scripts) will be required for the optimum operation of your website. Server-side languages enable dynamic websites. Perl is the most popular scripting language, but others such as PHP, JSP, Tcl, Python, Miva Empressa, server JavaScript, and even compiled C/C++ are some of the other server-side languages utilized in the operation of a dynamic website. Windows-based hosting services will usually offer ASP, VBScript, Jscript, PerlScript and ColdFusion.

Specific CGI (Common Gateway Interface) scripts (small but highly potent bits of computer code) are important for an email response form, a site search engine, and e-commerce support. It's a way to provide interactivity to web pages, such as the handling of the input from forms, like using specific code to enable information to be taken from a web-based form and sent to your email account. CGI scripts enable your website to accept credit card orders, and track everything a visitor does once he gets on your site. It can also automate otherwise tedious processes such as signing customers up if you have a "registered users only" section on your website.

You will find that all web-hosting services are not constant regarding CGI script policy. Some services:

* Offer libraries of CGI scripts (CGI-bin directory) for your use.
* Offer libraries of CGI scripts (CGI-bin directory) for your use and let you install your own CGI scripts (your own CGI-bin directory).
* Don't have a library but allow you to install your own CGI.
* Don't allow you to add any CGIs.

Find a web-hosting service that supports the server-side language you need to operate your website and, at a minimum, allows you to install them in your own CGI-bin directory.

Note: You will need to reference server-side scripting code that reside in a CGI-bin directory. This directory will include code such as a program that enables an email message to be sent out by a web page form. Thus, for most e-commerce sites, it is absolutely essential that you sign with a web-hosting service that allows you to have full control over your own CGI-bin.

Telnet Access

A shell account is a login account on a UNIX server and is based on a UNIX operating system. A telnet shell account allows you to interact with your site in a multitude of ways. Telnet is a common way to control web servers remotely. Think of it as if you were sitting behind the keyboard of the server. It lets you change, remove, and make new files, you can check your disk usage, clean out your server logs, read files, test scripts, and FTP files from other servers. With telnet you can change the permissions of a CGI file, change your password, create new directories, or just use programming tools such as C and Perl. That's the capability a telnet shell account gives you.

Telnet is a powerful tool. It is a terminal emulation program (a program that makes your computer respond like a keyboard) for TCP/IP networks such as the Internet. The Telnet program runs on your computer and connects it to a remote computer (your web server at your hosting service). You can then enter commands through the Telnet program and they will be executed as if you were entering them directly on the server console. This enables you to control the server and communicate with any other servers you might have at the hosting service.

The ability to modify files and directories is very useful for an active site. Windows95/98/NT comes equipped with a built in Telnet program. To start a Telnet session, you must log into a server by entering a valid username and password. Web-hosting services are split down the middle as to whether they will provide their clients a shell account or not (and with legitimate reasons — the most prominent being that Telnet access give hackers another avenue for attack). Without Telnet access, however, you can incur longer programming development time, which adds to your overall development costs. Also, without Telnet access, you can't compile programs written in C, C++, Perl, etc. Instead you must rely on the hosting service's technical support staff, which usually results in irritating and expensive delays.

So while it is relatively easy to find a web-hosting service that allows FTP access to a CGI-bin directory, it is more difficult to find one that allows Telnet access. Telnet access is not a do or die requirement. Still, if you find a web-hosting service that offers it, and you are satisfied with the service's security set-up, it is an added bonus.

E-Commerce

When considering a web-hosting service's e-commerce offerings, please follow the advice set out in Chapter 9. As a "quickie update," remember:

* Before putting your e-commerce in the hands of a hosting service, look closely at the critical components. The software should be intuitive with some kind of "wizard" to assist you with the set up, such as tools that provide category pages with links to individual items along with "Buy" links that can take the customer straight to the shopping cart.

* To look at the catalog builder. Can you create your product catalog offline and then post it to your website and link it to the database you use to track inventory?

* The web-hosting service that offers an e-commerce package also should be able to help you with setting up a merchant account — the agreement between your web-based business and your bank that allows you to take credit card orders.

* Do you need real-time credit card authorization (see Chapter 3)? If so, this will be an extra cost.

* Look at how the web host handles security for its clients' order processing. Are all credit card transactions conducted over a Secure Socket Layer (SSL) server? If not, what security does your web-hosting service offer for order processing, especially credit card data? (See Chapter 7 for a detail discussion of security considerations.)

Technical Support

First-rate technical support should be your primary consideration, especially if you're not a technical wizard. The Internet has no down time. Your server must be available 24 hours a day, 7 days a week. In the event of problems, immediate contact and assistance from your web-hosting service's technical staff is mandatory. Look for a hosting service that provides automatic monitoring of your website. This way, if there is a problem with the site they will be able to respond to it immediately. Ask if the web-hosting service's stated "24x7" service policy means that server support, server monitoring, and server availability are all covered.

But still, that doesn't answer the question of just the quality of the hosting service's technical support. Customer support can make or break your hosting experience, so test their support before making a commitment. If the host has a 24/7 live chat, check it out. The best and easiest way to check out a potential hosting service's technical support is to check out the service's procedures for handling problems with your server. Try to determine how much help they actually provide individual clients. Ask:

* What type of technical support is available?
* Is 24-hour tech support available?
* Is there a separate tech support line for hosting customers?
* What is the average hold time before a *technician* picks up the phone to serve you and what is the average time for response to a client's inquiry?
* Are technical support calls toll-free or a local call?
* What is their time limit on response to customer inquiries?

Make sure you get thorough answers to your questions. Then, actually check out the potential hosting service's technical support. Here's how:

* Place a few telephone calls to technical support. Place calls at all hours of the day and night, and on different days of the week.
* With each call, ask the technical support personnel questions that would represent some problems you might incur while using the hosting service.
* Send them a question by email and see how long they take to respond.
* Check out the quality of their online documentation.

If you get the right answers stated in language you understand (i.e., not technospeak), and you didn't have to let the phone ring a 100 times or hold for 15 or more minutes, you've hit pay dirt.

Web-hosting services handle technical support in a variety of ways. Some hosting services have a policy of handling all their support through email. The problem with this is that it's too easy for them to ignore email or to delay responses. You will find that 75% of the time you will need to talk to someone.

A few low-cost web-hosting companies have telephone support, although many will not offer toll-free numbers; but at least you get to talk to a "live person." Other hosting services, some with toll free numbers, charge for telephone-based technical support. Just be sure you know where you stand before signing on that dotted line.

If you need 24-hour technical support — and larger companies and high-traffic websites do — then expect to pay substantially more. People are much more expensive than machines.

The best practice is to find others using the service and ask them about their experiences.

Value-added Services

All quality web-hosting services have a library of scripts that you can use to add forms, guestbooks, forms, statistics, and so forth to your site. Your web-hosting service also should provide e-commerce capabilities, e.g. shopping cart software, merchant account setup support, real-time processing availability, and more. Furthermore, some e-commerce operations may need support for Java, Shockwave, Cybercash, Real Audio, Real Video, VRML, secure transactions, and other utilities. It is important that the technical staff of your chosen web-hosting service be familiar with the applications you plan to use. For example, if you're planning to use an application that requires special setup parameters, make sure your potential host is familiar with the application. If not, you just might end up spending an enormous amount of time trying to figure out how to configure it, or perhaps never figuring it out!

Site Administration

Quite a few quality web-hosting services offer web-based site administration. While not an essential requirement, the proper web-based administration tools allow you to do things such as easily set up POP accounts and configure autoresponders through your web browser (i.e. the tools are basically all point and click). Even if the site administration tools are not web-based, you will still need to update your pages, manage files, collect orders, retrieve data from forms, get statistics, make counters, and perform other housekeeping chores to your site. So find out how secure and user-friendly is the software that you'll be using to do these things.

Also ask the web-hosting service:

* Can you make changes anytime you want?
* Do these changes need to be audited?
* Can you place custom ASP scripts on the server?

Remember that your website is a direct reflection of you and your web-based business. As such, you want to have sufficient control of your website so as to portray your business in the best "light" possible.

Flexibility

As with everything in life, flexibility is a nice perk. Ask if the web-hosting service will accept special requests or instructions.

Another concern is the ability to upgrade services. As your web-based business grows so will your hosting needs. Ask if you can start out with an economy package and then upgrade as your needs and budget increase. Also check out how much it will cost for you to add more disk space, transfer more data, create more email accounts, and so on.

Security

Security is of paramount importance. As such, find out what security features your web-hosting service offers or supports. Although many web-hosting services claim to be secure, when closely examined they fall far short of their claim. Find out if your web-hosting service can actually protect your data from the growing menace of outside threats and hackers. For example, the hosting service should have an expert security staff on call to dispose of any potential threats.

Some key security and reliability issues you'll want to cover with a web-hosting service include:

* What kind of Internet firewall does the web-hosting service have in place to keep uninvited visitors out of its servers?
* How often does it conduct security audits and what other proactive steps does it take to address potential security holes?
* How are hackers kept out?
* Are back-ups performed daily to ensure data is never lost?
* Are all servers on an uninterruptible power supply (UPS) so data is always available even when there is a power outage?
* Since you will be taking orders on-line and perhaps transferring sensitive information, you'll need a secure server (often referred to as an SSL server — a Secure Sockets Layer server). For instance, credit-card information typed into a form will be encrypted before being sent to your web server. Some web-hosting services charge an additional fee to use their secure server, so ask. Look for hosting services that support transaction encryption standards like SSL and SET (See Chapter 7). You don't have to have a secure server to take orders on-line, but many people won't place orders unless you do.
* How does the web-hosting service get sensitive information from your web server to your back-office in a secure manner?
* Does the hosting service use redundant connections so your customers can access your website even if a line goes down or is cut.
* Is the site physically secured so that only authorized personnel have physical access to servers?

Finding a web-hosting service that will maintain the integrity of your website is nearly as important as maintaining its availability. Having your server "hacked" is not something you want to experience so find out if the web-hosting service has taken the necessary steps to secure your website. In addition, ask about partitioning of users on shared servers, ability to encrypt user access, and how the software security of the server is set up to prevent unauthorized access.

Email

Almost all web-hosting services provide their client websites with at least one email account, although most will provide several accounts. However, some hosting services charge extra for an email account, so ask. Also determine if there is POP access to the emails since some web hosting services only offer web access to the email accounts.

Ask the web-hosting service how many email addresses you are allowed since some services allow you to set up multiple "aliases" such aswebmaster@yourcompany.com, sales@yourcompany.com or info@yourcompany.com. Another feature web-based businesses might find useful is the ability for different aliases to be forwarded to more than one email address. For example, you might have stores in many different locales, with email aliases for each of them.

Another mail forwarding feature offered by some web-hosting companies is the ability to automatically define certain types of incoming email messages that are to be forwarded somewhere else. For instance, messages to marketing@yourcompany.com could be forwarded to marketing@acmemarketing.com. Remember to ask if there is a limit on how many accounts can be forwarded.

Another feature you might need is POP (Post Office Protocol) email boxes on your web-hosting site, although for some smaller web-based businesses, the POP email box you have with your local access ISP probably covers your needs.

Larger web-based businesses will need a minimum of 10 POP mailboxes. In addition to unlimited email aliasing most will want the option of designating a catchall POP account so that any mail sent to anaddresseeunknown@yourcompany.com will be forwarded automatically to the catchall account when a customer incorrectly addresses an email message.

Autoresponders

You will also want unlimited autoresponders. They undoubtedly will be an important part of your marketing strategy.

A mail responder, or autoresponder, is a program that automatically responds to incoming mail sent to certain addresses. Autoresponders let you to specify email addresses that allow for automatic and immediate posting of a pre-determined reply. For example, if someone sends email to sales@yourcompany.com, a brochure-type email message can be sent back.

Among the many useful attributes of a good autoresponder are that it allows you to:

* Save the incoming message.
* Copy the email address from the incoming message and put it in a text file, which is important since that allows you to easily collect email addresses.
* Quote the incoming message in the autoresponse.

Mailing List Management

A mailing list is a discussion group based on the email system. You can also use a mailing-list program to distribute a newsletter. A web-hosting service that offers the option of mailing list software such as Majordomo, is a bonus for all web-based business and a must have for others. Ask your web-hosting service if a mailing list option is available and if so, the cost, how many mailing lists are permitted per site, and if there is a limit on the number of members per list.

Online Promotion

Some nice perks that numerous web-hosting services offer involve help in promoting your website. For instance, some help with the registration of your website with the leading search engines and directories. Others may offer to help promote your site. And some web hosts will even assist your website with online advertising.

Access/Traffic Reports

The reason you are putting up a website is to attract customers. Find out what access/traffic reporting services your web-hosting service provides. For instance, do they provide access to raw logs and statistical reports to help you determine who is visiting your e-commerce site? It is vital that any web-based business knows:

* How often its website is being accessed.
* What web pages are most popular.
* Where its customers are coming from (i.e., search engine referrals).
* How the customers are viewing the website (i.e. browser type, connection speed).

If you have access to the raw server logs, you have the tools necessary to make sure that the site is functioning properly.

A good web hosting service also will provide software that tracks all the traffic to and from your site with the necessary statistical data on that traffic (your customers). Some hosting services will automatically email reports to you at pre-determined intervals. There are hosting services that even provide nice graphics showing the break down of a number of detailed categories with charts and numbers.

Even if you are satisfied with the access/traffic reports provided by the web hosting service, you still need to find out a bit more. For instance:

* Ask the service to provide sample access/traffic reports so you can determine if the information provided suits your needs.
* Ask if the reports are accessible online, at anytime.
* Find out if the reports held in archive for later reference and for how long.
* Ask how often the reports are compiled.

Multiple Domain Names

Ask your web-hosting service if it charges a fee for extra domains. Extra domains are very useful. Your primary domain is for your main web-based business but you could use additional domain names to promote specific products. You can structure this in various ways. For instance, all the domains can point to the same directory; you can have separate directories for each domain; or a specific group of domains pointing to a single directory. Web-hosting services have a variety of billing methods for this option. Some charge an additional fee for each extra domain; others allow a set number of domains and charging for any over that limit; still others charge nothing at all for domains. You have to ask.

File Transfer Protocol (FTP)

You don't want to be restricted in any way when it comes to updating your web pages. Look for a web-hosting service that will provide unlimited FTP access since that is how you upload your web pages to your web server. Most web-hosting companies provide this service.

Another type of FTP service you will need, if your website offers software downloads, is anonymous FTP (sometimes referred to as "public FTP"). This is different from FTP access to your website. Although it is possible to transfer files directly from your website, customers that don't have decent web access will want to use the anonymous FTP option. Another feature of some anonymous FTPs is that an interrupted download can be resumed, i.e., if a file transfer stops somewhere in the middle due to an ISP or telephone line problem, the customer can continue the transfer where it stopped when he/she is able to get back online.

Server Performance

Another must is a high performance web server. Do not compromise on this issue.

Find a web-hosting service that offers high quality server hardware with fast CPUs, a lot of memory (the more the better), high-speed disk drives, and redundant T3 connections or better. Ask the brand names of the servers, the operating system loaded thereon, and the web server software provided. Also determine how often the web-hosting service updates and replaces their servers' hardware and software.

Test the servers. Measure the speed of access for all facets of access (network speed, DNS lookup, connect time, download time, and absence of timeouts). Be sure you run tests during both peak and non-peak times.

One of the easiest ways to test your web-hosting service's servers is to use Net Mechanic's Server Start (www.netmechanic.com). This is a nice free tool to use to measure a server's access speeds. This product allows you to watch server perform-ance for 8 hours and will send you an email report.

Password Protected Area

Portions of your website can be password protected so that access is limited authorized personnel only. Password protected areas is one of the best ways to keep unwanted cretins out of your website.

Unfortunately, these protected areas can be problematic to set-up. Most web-host-ing services will either assist in the creation of these protected areas or take care of all of the technical aspects of setting up your password protected area(s).

Backups (Server and Power)

Information on your servers will be in a constant state of change. Making daily back-ups is essential in case of a major server crash. Ask the web-hosting service if it pro-vides off-site storage of the backup tapes. If you also have your database on the web host's servers, talk to them about their procedure for backup data or if they can help you backup the data yourself.

Another important issue is a power backup system. Most web-hosting services do this as a matter of course, but it doesn't hurt to check that your web host provides UPS and generator power backup.

Redundancy

The ability to be up and running no matter what is of critical importance. There are many elements to redundancy — see Chapter 5 for the full picture. However, a good web-hosting service will have redundant Internet connections, which are essential for providing uninterrupted service to your customers. With only one connection point, if the service's pipe to the Internet goes down, then your website is down.

Scalability

The best thing that can happen is that your web-based business experiences a giant spurt of growth due to its popularity. However, can your web-hosting service handle such growth? If you feel that increased traffic on your website may become an issue, find a hosting service that quickly can add bandwidth and processing power to meet any level of demand. A good quality web-hosting service will automatically moni-tor on a 24x7 basis, both the servers and the Internet connections so that it can respond appropriately to any traffic or congestion problems, without any action on your part.

Databases

As discussed in Chapter 8, databases have their own issues. If you use a database to store your data, or your website is database driven, you need a web hosting service that offers database servers. Most Windows-based services will offer Microsoft Access and/or Microsoft SQL Server. Most Unix-based services will offer some form of SQL (MySQL, PostgreSQL MS SQL Server) and/or Oracle. Be very careful to place your website on a web-hosting service that offers an operating system compatible with the system you use to maintain your database.

Other Items

Other features to consider when looking for a web-hosting service include:

* A static IP address. If not, go elsewhere.
* Provide sub-domains for creating sub-sections to your website without the necessity of using directories in URLs.
* A control panel for graphical access to your account settings.
* Guaranteed uptime that includes a refund for times when your host is down.
* Find out if the hosting service has larger plans available so you won't have to deal with the task of changing hosts if your website requires more space, features, etc. in the future.
* Is the web hosting service offering any special deals, discounts, promotions?
* If you want a specific version (or the latest version) of a web server, shopping cart program, database, and/or server-side language, ask if that version is available, don't assume it will be the version provided.

⊙ SERVICE LEVEL AGREEMENTS

Before signing on the dotted line, carefully read the web hosting service's "Terms of Use" and "Acceptable Use Policies." Many hosting services put important qualifying statements in those documents. For example, this is where content limitations (if you have a controversial site) are usually stated. That is also where you will find statements qualifying the host's uptime guarantee (are they responsible only when the server is down, or also when their connecting network is down?), and limitations on the so-called "unlimited" plans. Also look for "catches" such as daily bandwidth measuring, sale of personal information, and limits on the size of your database.

Go with a hosting service that offers, at a minimum, a 30-day money-back guarantee, which should be posted in easy view on their website. However, even with a money-back guarantee, the setup fee might not be refundable, so check.

If everything is copasetic, the last item for discussion is the service level agreement. While you might attempt negotiation of a service agreement, you probably won't get very far. So be realistic; a web hosting service will not guarantee to provide 100% uptime, nor will it offer more than pro rata compensation for down time or reimbursement of paid service fees. That being said, most of the down time your site experiences will probably be a direct result of something that is done from your end and not from service outages attributable to the hosting service.

✪ CONCLUSION

Choosing a web-hosting service is not a task to be taken lightly. Finding a hosting service that provides a plan with the features your website needs for optimal operation, at a price you can afford, could try Job's patience.

When selecting a web-hosting service, your first decision is whether you need Windows or Unix-based hosting service, then how much disk space your website will need and how much data transfer you expect. Most of you also will need a web host that provides database access and SSL servers. Furthermore, some e-commerce operators will need to consider whether the server-side languages that the web-hosting service provides meet their site's needs.

Finally, carefully read the hosting service's "Terms of Use" and "Acceptable Use Polices" prior to entering into a hosting agreement. Be sure you understand these documents because that is where, for example, many services place qualifying statements concerning uptime guarantees and limitations on their "unlimited" plans.

Chapter 15

Search Engines and Directories

Whenever you see a successful business,
someone once made a courageous decision.

—Peter Drucker

You've designed a wonderful website, but you may be disappointed with the traffic and sales your site is generating, so you look for a solution. As you research the problem, you'll find the majority of a website's traffic comes from search engine and directory referrals — your website may not be getting these referrals. The reason may be that you have not designed your pages correctly for search engine placement or you have not submitted your website correctly with the various search engines and directories (or a little of both).

We all know that it's sometimes difficult to find what you're looking for on the World Wide Web. Even when using a search engine or directory service you still need a fairly specific search — if you don't want a million returns. Your potential customers are in the same predicament. Take the time and make the effort to guarantee that your website is listed well for the keywords that are important to you. Most search engines use spiders to crawl the Web in search for new pages to add to the services database or index. You can wait for one of these spiders to locate your site, go to the search engines' websites and tell them about your new website, or use one of the many site submission services.

Let's examine the three different methods used to search the Web: crawler-based search engines (e.g. AlltheWeb), directories (e.g. the Open Directory Project), and a hybrid service (e.g. Yahoo!).

✪ SEARCH ENGINES

The vast majority of search engines use spiders. These ingenious software programs have only one task — to crawl the Web 24 hours a day, finding and indexing web pages. These spiders (also called "bots" or "crawlers") visit a web page, read it, and then follow links to other pages within the site. The spider revisits websites on a regular basis (e.g. every month or two) to look for changes.

Everything a spider finds is put in the search engine's index. The index, sometimes called the catalog, is a giant database that contains a copy of every web page that the spiders find. When a web page changes, the index is updated with the new information, but it doesn't happen immediately — it can take a while (as much as six weeks) for new pages or changes to be added to an index. So although a web page may have been "spidered" it may not have been "indexed," and until such indexing occurs, that new page is not available to those searching with that search engine.

Search engines do not index the entire Web (although it may seem like they do). Most also don't include dynamically created web pages like library web catalogs or other data behind CGI-walls. And none index the entirety of every website, nor do they share a common search language (i.e. the algorithm used to determine what is searched and ranking of the searched item varies depending on the search engine). However, there are three important elements that are common to all search engines:

1. The database operates on the same principles as your website's database. The database consists of indexed descriptions of web pages including a link list with a small description for each link. When a search request is received from a surfer, these databases utilize special search algorithms, using keywords, to find needed web pages.

2. Search engines give each page they find a ranking as to the quality of the match to the surfer's search query. Relevant scores reflect the number of times a search term appears, if it appears in the title, if it appears at the beginning of the page or HTML tags, and if all the search terms are near each other. Some engines allow the user to control the relevance score by giving a different weight to each search word. A search term used too many times within a page can be considered web spamming (for which search engines penalize) so don't overdo the use of a keyword or phrase on a page (don't exceed the 15-25 count range).

3. Each search engine has its own peculiar ranking method. For example, if there are no links to other sites or pages within a website (a single page website) some search engines will not list that website.

SPIDERS

A spider is an automated software program designed by search engines to follow hyperlinks throughout a website, retrieving and indexing pages in order to document the site for searching purposes. But what should concern a website designer is a spider's nuances — a spider determines relevancy, i.e. if someone searches for "beeswax candles," the search results will be only those web pages that contain the words beeswax candles. That is simple enough, but suppose there are more than one website with the term "beeswax candles"? Search engine results are presented in descending order of relevancy to the search term that was used. Relevancy determines which results will be presented first, and which second, and on and on. The spider's job is to work out which page is most relevant to the term "beeswax candles" and which is the least relevant.

Spiders calculate relevancy based on four factors: repetition, prominence, emphasis, and link popularity. Let's examine each of these more closely.

Repetition. This is simply the number of times a word is repeated on the page. The more often it is repeated the greater is its relevancy to the page. But resist the temptation to simply repeat the "keyword" over and over again because spiders are programmed to de-list a web page if there are too many repetitions.

Prominence. This is where keywords appear within the website. Originally all a spider looked for was the "keyword" meta tag, but not any longer. Now they look in keyword meta tags, description meta tags, alternative text tags (on images), page titles, body text, and link text.

Emphasis. The number of times a search term appears, and whether it appears in the title and at the beginning of the page or HTML tags; and if all the search terms are near each other. Note: some engines allow the user to control the relevance score by giving a different weight to each search word, others won't.

Link popularity. This is the number of third party sites that are linked to a website. Each link is regarded as a "vote" for the site. But to complicate matters, the "votes" carry greater weight if the linked website is one that the spider recognizes as having a similar theme as the web page it is crawling.

While all crawler-based search engines operate basically the same way, there are differences, which is one of the reasons the same search on different search engines produce different results. For instance, when spiders submit their results for indexing, either the data is placed directly into the search engine's index, or it is vetted by humans prior to indexing. Once your pages are added to the relevant indexes, your potential customers can search using various keywords and phrases to find pages that best match their search criteria. Search engines also use software programs to sift through the millions of pages recorded in the index to find matches when a search term is entered. These software programs rank the web pages contained in the index — how the programs perform the ranking task is kept a closely guarded secret.

Appearing within the first 20 returns of any relevant search is critical to driving customers to your website. But since each search engine uses its own special "magic" to determine the rank or position of importance of each individual website, the exact rules that the search engine uses to rank pages for relevance are generally tough to ascertain (and they change often).

✪ DIRECTORIES

For the most part, directories such as Yahoo!, the Open Directory Project (http://dmoz.org/), IDB/Internet Directory for Botany (www.botany.net/IDB), Writers.net, the Middleeastdirectory.com, Questia.com (an online library), etc., use real people to compile their information and then supplement that information with a search engine. A directory-based search returns matches only when one is found either in a description submitted by the website owner or contained in a review.

Since directories are dependent upon humans for their listings, a website operator is required to submit a short description of his or her website. Then depending on certain (unnamed) criteria, an editor may then visit the submitted website and write a review. (A well designed website, with good content, is more likely to be chosen for review than a badly designed website or one with poor content.)

Changing your web pages has no effect on a directory listing. What will improve a listing with a search engine will not serve to improve a directory listing.

Note: The average time between your original submission to a search engine or directory and getting it into that service's database is about five to eight weeks.

✪ HYBRID SERVICES

These services present both directory-based and crawler-based results, although most hybrid services favor one type of listing over another. MSN Search, for example, typically will present listings from the LookSmart directory before it presents crawler-based results provided by Inktomi. Yahoo! uses its own directory service, which is supplemented with crawler-based results provided by Google (that may have changed by the time you read this book).

Designing for Search Engines

But, you might ask, how do crawler-based search engines go about determining relevancy, when confronted with hundreds of millions of web pages to sort through? They use algorithms, i.e. a set of rules that govern their spiders' crawling techniques, indexing techniques and ranking within the list of returns of a specific search term. Although exactly how a search engine's algorithm works is a closely kept secret, all major search engines follow the same, general rules.

The remainder of this chapter, hopefully, will help you to understand how to design your web pages so that your website will get the proper search engine and directory rankings that it needs to be successful.

When someone queries a search engine for a keyword related to your site's products/services, does your web page appear in the top 20 matches, or does your competition's? If your web pages aren't listed within the first two or three pages of results, you lose. To avoid such a circumstance, when designing your new website, take into consideration the inner workings of search engines. If you ignore the criteria necessary for optimal placement by search engines, your website will miss out on traffic that it would otherwise have received *if* your website had been designed with search engine placement as one of its design criteria.

Note: The three most popular search portals are Google, AlltheWeb, and Yahoo!. Trailing behind these giants are MSN Search, AOL Search, Askjeeves, and HotBot. All of these search portals in one way or another use spiders to crawl or search the Internet. Humans then search through the results in an effort to optimize the search engine's database.

Mergers and acquisitions are changing the search portal landscape. As of mid-October 2003, Yahoo! owns AltaVista, AlltheWeb, Inktomi, and Overture. At this writing, AltaVista and AlltheWeb continue to be available at their historic locations; however, they may share the same underlying database very soon. It is noted that Inktomi remains the back-end search engine at MSN Search and is still available at HotBot.

As discussed previously, a spider is a small program that gives weight to the placement and frequency of words, and uses ranking algorithms during the search process. And as explained in the "Spider" text box, while location and frequency of keywords on a web page is generally given the most weight, related words and word relevance along with other criteria, such as descriptive five word or so page titles, body copy, placement of keywords, and meta tags within your HTML code, etc. all play a role in how a search engine ranks your web pages.

Here is an illustration of how your website's design influences a web page's ranking:

Say that a potential customer types in "antiquarian books." That customer wants to find websites that have content about and/or sell antiquarian books. Since the search engine assumes the same thing, the results will be top heavy with web pages having that search term appearing in their HTML title tag. It assumes those web pages are more relevant than those without the term in their title tag. But search engines don't stop there. They also check to see if the words "antiquarian books" appear near the top of a web page, such as in the headline or in the first few paragraphs of text because it is assumed that any page relevant to the topic will mention those words somewhere near the top of the page.

Now let's consider frequency using the same scenario. A search engine also analyzes how often the keywords "antiquarian" and "books" appear in relation to other words in a web page. Those with a higher frequency are often deemed more relevant than other web pages.

So even though search engines vary on how they rank websites, every web page should include:

* Page <TITLE> tag.
* Keyword meta tag which is more than one word.
* Description meta tag.
* <!— comments tags —>.
* First 25 words (or 255 characters) of text.
* NO FRAMES tag.
* Hidden FORM tag.
* HTML tags.
* <ALT> tags.

Let's look a bit closer at each of these elements:

Title: The title you choose will be the most important decision you make affecting search engine ranking and listing. There is no specific science to it — just make it simple. Look at the web page and the first five or so descriptive words that come to mind can be the title. Another way to look it — think of your title as a catchy headline for an ad.

Text: When it comes to the text of a submitted web page the search engines vary their indexing procedure. While some will index the text of a submitted page others will only take into account the first 25 words (or 255 characters) of a submitted page (25/255 rule). So, write the text of a submitted page using the important keywords more than once in the first 25 words.

Something else you can do is to create at the top of a submitted web page, a transparent gif image that is one pixel in size and inside the ALT tag insert a description of the page using the 25/255 rule.

Meta Tags: They are indispensable tools in your battle for search engine ranking. Put them, along with keywords relevant to each specific page, on each page of your website.

When we discuss meta tags in this chapter we are discussing only description and keyword tags. A description meta tag is exactly what it sounds like — it gives a description of a web page for the search engine summary. A keyword meta tag is again exactly what it states — it gives keywords (which should never be fewer than two words) for the search engine to associate with a specific web page. These meta tags, which go inside the header tags, are crucial for optimal search engine indexing. Your meta tags should reflect the content of the first couple of sentences of the main body. It is important that you make certain that the words you use in your keyword tags are words that someone would type in to find your website.

Keyword hints:

* Keywords are target words that will drive people to your website.
* When choosing your keywords, always use the plural of the word. Searching for "car" with find sites with "cars" in their keywords but searching for "cars" will not find sites with only the singular "car" in their keywords.
* Almost any site on the Web could use "web," "internet," "net," or "services," as a descriptive keyword. Don't! Using these and other like words to target potential customers is fruitless and most of the spiders actively ignore common words such as these.
* Include incorrect spellings of keywords that are routinely misspelled. For example, the word "accommodations" is commonly misspelled as "accomodations" so include both in your keywords.

An example:
```
<HEAD>
<TITLE>Best Online Widget Store in the Universe</TITLE>
<META name="description" content="An online store with all the Widgets you would ever want.">
<META name="keywords" content="widgets, widget accessories, widget howto, widget books, widget articles, widget technical papers, widget software, working with widgets, designing with widgets,">
</HEAD>
```

For guidelines on what you should do with meta tags, go to a search engine, say AlltheWeb, search for a term or word that you hope someone would use to find your website. Then go to the top ranked websites and use the "view source" feature of your browser to see what kind of meta tags each of these sites use. Study them and understand their relationship to the web page, then use this information when you are composing your own meta tags.

Keywords: There are two ways to approach keywords: A blanket strategy and a targeted strategy. When you use a large list of keywords, your pages will be found by a variety of surfers using a extensive range of search strings, but your web pages will not, in all probability, be among the top ranking pages — this is the blanket strategy. When you use a limited number of keywords, the density of these few keywords increase and therefore put them higher up the list — the targeted strategy.

If you have a website that offers either a limited number of products/services or products/services that can be adequately covered with a short keyword list then the target strategy is for you. In other words, you're confident that potential customers will search for those specific words above all others. However, if you have a wide variety of products/services on your website (such as Drugstore.com or Outpost.com), then you might use the blanket method. Or consider the doorway page, mentioned later in this section.

Keyword Mix: Pay attention to your keyword mix. Keyword density (the ratio of a keyword or keyphrase to the total words [depth] on a page) is the factor that search engines most consider when assigning relevancy ratings to web pages. Achieving the right keyword or keyphrase mix has become almost a science. Some search engines look for various combinations of keyword density, i.e., the number of keywords versus total word count must be within a certain range, and to complicate matters, they assign different "weights" to components such as Title, Meta Tags, Links, Body Text, Headline, etc.

Link popularity. This refers both to the number of similar websites you've placed links to within your web pages *and* the number of websites that have links that point back to your e-commerce site. Your links to other websites must be on relevant pages — that is pages that have as much to do with the common theme of your website as possible, and that are not just a page full of links. Pages that are full of links are commonly referred to as "link farms" and are ignored by spiders.

Search engines view a website with a large number of incoming links (i.e. other websites that have links to your website) as an important or popular website. Thus, according to search engines, a website with lots of links leading to it generally implies that the website is a valued one and the search engine's database would not be complete without it. Link popularity is vital if your site is to achieve a high search engine placement ranking.

Use Optimization Tools

If your keyword density is too low, your page will not be rated high enough in relevancy and, conversely, if too high, then your site may be penalized for "keyword stuffing." I know this sounds very complicated but there is a way to get help — keyword optimizing tools such as the GRSoftware's Keyword Density Analyzer (www.grsoft-ware.net) and the Webpositioning Gold at (www.website-promoters.com). Or check out Keyworddensity.com, which provides free online analysis of any web page. Another option is to use a service like Abalone Designs (www.abalone.ca), Dragonfly Design (http://dragonfly-design.com/special-offers.html.) or etrafficjams.com. All three provide free website analyses as a marketing tool.

Many of you will decide to purchase search engine optimization tools because such tools can help you to get the desired results from search engine submissions, by providing you with the means to tweak your website so that it works with visiting search engine spiders and their complex, math-based formulas used to rank websites. Here is what to look for when selecting a search engine optimization tool:

* Good documentation. You want a product that provides a clear overview of the different steps you need to take to prepare and then submit your website to the various search engines and directories.

* An intuitive interface that provides ease of use. When you buy a search engine optimization tool, you don't want to spend a lot of time learning how to use the program.

* Submission options. Look for a product that will let you choose the number of pages that can be submitted to the search engines. With most search engines, it is better to submit all of your web pages individually, but not all tools allow this type of submission.

* A viable search engine list. Watch out for search submission products that say they will submit your site to thousands of search engines. While this might help drive traffic to your website (there are thousands of micro search engines), only a small portion will help you increase the number of relevant visits to your website. Submitting your site to Lawcrawler, or some other specialty search engine unrelated to your website's content, will not have much value. The best search engine tools focus on the major search engines and directories. Then add regional and specialty engines.

* Keyword selection. Choose a tool that guarantees that its keyword selection tools are based on an analysis of millions of words and phrases entered in search queries daily.

* Page analysis. Look for a product that offers page analysis, i.e. there are tools to examine each web page from a search engine's viewpoint. The tool should check for standard techniques and elements that will improve your ranking for your selected keyword phrases. A list of problem areas and how to correct them should be presented after the analysis is complete.

* Select specific search engines and directories. Find an optimization tool that

doesn't provide an "all-or-nothing" submission option. Instead, choose a search engine optimization product that allows you to select the search engines and directories to which you want to submit your website pages.

* Tracking and ongoing analysis. Look for a product that enables you to assess whether your web pages have been listed in the index of all submitted search engines and directories and how they are ranked in the various indexes. Products that provide tracking and ongoing analysis tools will allow you to perform such analysis.

* Money-back guarantee. Don't spend your money on a optimization tool that won't provide a money-back guarantee. Many search engines and directories are now very selective about automated submission sources (this is what all search engine optimization tools do). Find a product that guarantees listings in the search engines and directories on its list.

Most search engine optimization tools are relatively inexpensive. To give you a sampling of what's available, check out:

* Axandra/Voget Selbach Enterprises GmbH's Internet Business Promoter (IBP) product (www.axandra). The cost ranges from free to $350.00.

* Microsoft's bCentral (www.bcentral.com). The cost for the annual plan is $79.

* Netmechanic.com's search engine optimization tools are available on an annual plan basis. The cost per URL subscription is $49.

* Websiteceo.com offers four editions with different pricing options. The cost ranges from free to $495.

Note: *By submitting your website to a search engine you speed up the spidering process. For after you have submitted your URL to a search engine, it sends a spider to "investigate" your new website. But you should also be aware that the information the spider returns many times will be exactly what appears in the "results page" of the search engine.*

You Can Do More

Now let's examine some other steps you can take that *might* improved your web page ranking.

Cloaking, or "stealth scripts." These pages make use of software to serve one page to surfers and a different page to search engines. There are even sophisticated scripts that can serve a different page to each search engine, allowing you to customize a page for specific engines. A secondary reason for cloaking your pages is to protect your highly optimized page from code thieves. All a thief can steal is your "surfer" page.

It is imperative that your search engine pages (i.e. cloaked pages) represent your surfer pages fairly and accurately. Search engines once threatened to ban cloaked sites.

However, most engines have admitted, reluctantly, that they will not take any action unless the surfer page is on a different subject than the search engine page. But if you use keywords and phrases that are not related to your actual content, you risk having your entire website banned from most, if not all, search engine indexes.

Doorway Pages: Search engines do a poor job of indexing and scoring web pages that use dynamically generated pages, frames, or Java Script. This is where doorway pages, entry or bridge pages (they all mean the same thing), come into play.

Doorway pages also can be used to create alternate entrances to your website so as to target a specific search engine with a page designed to deal with that search engine's criteria. Although doorway pages should be designed carefully to target specific keywords for individual search engines, they should also provide customer-centric information and point the customer to the "guts" of your website and have the same look and feel as the rest of your website.

Doorway pages can help you to obtain a high ranking using the unique ranking algorithms of each search engine. A web page that will be highly ranked by one search engine may not fare as well on another. You will want to cater to the top search engines, which at this writing are: Google, AlltheWeb, Yahoo!-owned Inktomi, and Teoma (Ask Jeeves). Ideally, you need one gateway page for each of these six search engines, and for each keyword or keyword phrase that a potential customer might use to find your site. For example, if you anticipate that normally customers will search for your website using one of five different keywords/phrases, then you'll need 30 doorway pages — 6 search engines times 5 keywords/phrases.

There are three types of doorway pages:

* A page that invites the customer to continue on to your website's home page (at the same time it provides the specific search engine with a page that it will find highly relevant).

* A page that is semi-invisible to the customer through the use of a Java script redirection technique. The page that is submitted to the search engines is stripped down to the minimum so that the search engine finds it highly relevant but a customer will only see the page as a "flash" before the real page is presented to the browser. There are two problems with this method: If your customers' browser is not enabled for Java script, they will see the unattractive stripped down page, and some search engines find the "redirection" code and then downgrade the relevancy of the page.

* A doorway page that is completely invisible to your customers. This is where software like iPush.net — an IP based delivery and cloaking system designed to help

you get the best possible search engine results and keep them. An oversimplification of this software is that it "cloaks" the doorway pages so that only the spiders from a specific search engine sees a specific page. This allows you to:

- Never worry about your visitors seeing your doorway pages, since they won't see them, you can design them strictly based on a particular engine's algorithm. They may make no sense at all, but the engine they were designed for will love them.
- Never worry about someone stealing your rankings because outsiders won't be able to see the HTML code that got the ranking.
- Design your regular pages freely without fear of hurting your rankings.
- Use multiple pages without worry of hitting any limit in any search engine's criteria (www.ipush.net for details).

As a side note: Yahoo! does not accept doorway pages.

To find all the nitty gritty details necessary to design effective doorway pages, search the Web with one of the many search engines using the keywords "doorway pages." Or visit Spider-food.net for extensive information on doorway pages and other search engine goodies. Another great site is Spiderhunter.com where you can find a tutorial on cloaking techniques and free cloaking script for your use, plus many other interesting details about spiders.

Submitting Your Web Pages

The best practice (there are exceptions) is to submit each page of your website, individually, to all of the search engines and directories. You may think that submitting each and every page of your website is not necessary since some pages may have, for example, investor information or contact information. But every page that is listed is like an entry in a drawing — the more entries, the more chances you have.

Use a search engine optimization tool (previously discussed) for the submission process, you can submit some of your web pages by hand to specific search engines and automate the rest using web-positioning software such as Submitta.com, the previously mentioned Webpositioning Gold(www.website-promoters.com), or check out one of the many website promotion services available online.

If you want to try submitting your web pages yourself here is a guide:

* Submit your main URL (i.e. http://yourdomainname.com) after you have finished designing your website.
* Submit other important pages in weekly intervals and in very small batches (no more

than 10 a day) since search engines are very sensitive to what they consider spamming.

* A large website should submit first its most important and customer-centric web pages, keyword-wise that is, since it is easy for a website with 200 or so pages to hit their page limit (usually 50 or so pages) with search engines.

* Once you have submitted all of your web pages, you need to not only re-submit each time you make substantial changes to a page but also, once every three or so months re-submit the pages following the procedure set out above.

* Pages that are generated "on the fly" usually will not be indexed so don't submit them.

* If you have pages with frames, don't submit them since most spiders will not crawl a page with frames (no matter what you might have read to the contrary).

* Test and check to see how your website rates with the search engines after your submission procedure is completed.

* Monitor your website listings regularly. Sometimes your listing can just disappear or some kind of error can cause the link to become bad, etc. When you find something wrong, re-submit that web page.

Pay attention to how your website is listed on a search engine. Does it identify what your website is and the products/services provided? To assure that your search engine listing provides the proper information needed by the surfing public, use your title tag (e.g. <title>Best Online Widget Store in the Universe </title>). Search engines then use as the descriptive paragraph one of the following, depending on the search engine: *either* your description tag (<META name="description" content="An online store with all the Widgets you would ever want.">) *or* the first 250 words (or so) of visible text on your site.

Although it will take a little effort, it is important that you balance these tags. In other words, sometimes what you need to put in as a title or descriptive tag (to get a high ranking with a particular search engine) will not help you in your quest to have potential customers easily find your page through commonly used keyword searches.

✪ DESIGNING FOR DIRECTORIES

Directories are differentiated from search engines in that real people, instead spiders and software, directories categorize the information contained in their indexes. Each directory has an in-depth submission form which the website owner must fill out and submit before the website can be listed in the directory.

When filling out a submission form, keep in mind that directories have real humans with inquiring minds working behind the scene. The better you present your information the better chance you have of getting your website listed in the directory.

Here are some tips to help you in the directory submission process:

* It is vital that you carefully consider and choose your 25-word description and category that you want your website to be listed under. Once it is submitted it's very difficult to change your mind later.

* You should endeavor to find the best category listing for your website. To do this, go to that directory, search for a word or phrase you hope would be most used to find your website. The resulting category is the category for which you should submit your website.

* Provide an accurate, concise description of your website — using 25 words or less is the key to success. It is interesting though that if your website has graphics or photos and you indicate that fact in the submission form, your website will have a slight edge over other similar sites without graphics or photos.

Remember, directories are compiled by thinking humans, so a good site, with good content, has a better chance of being listed than its lesser contemporaries.

✪ YAHOO!

Yahoo! is a directory that also integrates search engine results into its returns. This makes it a category all unto itself. Through an alliance with first Alta Vista and then with Hotbot and now with Google, when a surfer makes a search request, if no listing is found within Yahoo! then the search is automatically and transparently defaulted to the larger, spider-generated database. So you may think your website has a Yahoo! listing when you see your site in a Yahoo! results page, but that listing may have been generated from the search engine's larger database. This isn't the same as a true Yahoo! listing.

It is very difficult to get a good listing on Yahoo!, but it is well worth the effort since the site, as a whole, is the most popular site on the Web. Yahoo can generate much of your site's traffic — if you can obtain a successful listing.

✪ SPONSORED LISTINGS

To ensure that potential customers see your website listing when searching of specific terms, consider sponsored listings. If you can afford it, a sponsored listing will give your website top billing without the worry about whether your web pages are optimized for search engine spiders. Yahoo!, Google, and other search engines offer sponsored listings which enable your URL to be displayed in a highly visible ad space when the results of a specific search term is returned.

These services all work basically the same way. They have a method of determining the "known" number of searches performed based upon a certain keyword, such as "widget." Then, if a potential customer goes to a search engine where you purchased the keyword phrase "widget repair" (and maybe others, budget permitting) and types "widget repair" that potential customer would see not only the normal search result, but also your website's sponsored listing on their computer screen. If all available impressions were purchased, the potential customer would see your ad every time a search for "widget repair" was performed. If only a portion of the total impressions is purchased, the ad would rotate with other sponsored listings.

There are also dedicated paid listing services that resemble your average search engine or directory. You will find that potential customers use these services — just check out the popularity of Overture.com (formerly GoTo.com), findwhat.com, one-search.com, etc. If you are so inclined, a dedicated paid listing service may be just the ticket to driving more traffic to your website.

✪ CONCLUSION

Understanding the peculiarities of search engines and directories is a complex subject. I have tried to cover what is necessary to give your website the best chance for breaking into the coveted "top 20" inner sanctum. But for those willing to delve deeper, essential websites to bookmark include Searchenginewatch.com (also subscribe to their paid service), Positioned1.com, and Fantomaster.com. All are good quality websites with wonderful information; use them often.

Finally, don't obsess over your search engine and directory listing and ranking, just follow what is set out above, and then move on. If you don't get traffic from one certain search engine, yes, check it out, but if you can't figure out why, forget it and move on. There are many ways to market your new website, search engines and directories, although very important, are just one part of the package.

Chapter 16

Marketing Your Website

The market is a place set apart where men may deceive each other.

—Diogenes Laertius

Your new website is up and running, it's time to inform the purchasing public. Marketing a web-based business is just as important as marketing a brick-and-mortar business. Even though a marketing plan is probably one of the most important elements of a successful business, it is usually the least thought of aspect of any new business. Formulate a strategic marketing plan. A properly instituted marketing plan gives your website an acceptance level far above the average website by providing a rational direction for your marketing activities.

✪ THE MARKETING PLAN

Creating a good marketing plan is the best thing that you can do to help assure a new web-based business's growth. Your marketing plan should be your guide on which you base your marketing decisions and it will help to ensure that everyone involved in marketing your website works toward the same goals. A properly drawn up and instituted marketing plan not only provides a guide for the growth of your web-based business, but also how to spend your promotional dollars. Optimally a marketing plan and its budget (which is an integral part of your business's overall budget) should cover promotion and advertising for 6 to 12 months.

Don't forget to include in your plan's budget a sizable allotment for market research. What you know about your target market and the information gleaned from marketing research will give you the basis for your marketing strategy. Research is the only way you will know what is necessary to design your marketing plan to reach the 25-35% of your website's customers that are not brought to your site by search engines and directories. The plan should lay the groundwork for campaigns that will encourage customers to place an order, or to take some kind of action, that will allow you to respond — thereby establishing a relationship with another potential customer.

Preparation

Develop market objectives that are realistic and specific. If possible, hire consultants to assist you in identifying the available market, to understand who will be competing with you for that market share, and to formulate a realistic projection for your share of that market. Most of this information can be gathered from:

* Your internal records.
* Published market information from government statistics, chambers of commerce, newspapers, magazines, trade journals, banks, utility companies, city and county planning organizations, colleges and universities.
* Surveys, mail responses, telephone and personal interviews, opinion polls, market testing, and customer feedback.

Analyze your site from a promotional point of view. Then, with your marketing hat on, look at your site with a fresh eye and consider:

* What would a visitor consider as the main purpose of your website?
* Who would be interested in your website?
* How does your website stand out from other websites?

Now look at the competition:

* Are potential customers using the Web now?
* What are the strengths and weaknesses of the competition?
* What can be done to make the campaign's overall message superior to your competition's message?
* What message would be most effective in drawing potential customers to your website?
* What would cinch the conversion of potential customer into an actual customer?

This research will give you a good idea as to how you should go about reaching your current and potential customers — in other words, where you should spend your advertising dollars: banner ads, targeted opt-in email, newsletters (online or email), surveys, traditional advertising methods (print, radio, television), and incentives such as discounts, gift certificates and contests.

Now you have a good starting point for your strategic marketing campaign.

As you formulate your marketing plan consider:

Competitive Forces: Who are your major competitors now and who is likely to be your major competitors in the future? What response can you expect from those competitors to any change in your marketing strategy? How does the structure of the industry affect competitive forces in the industry?

Economic Forces: What is the general economic condition of the country or region where the majority of your customers reside (demographic research)? Are your consumers optimistic or pessimistic about the economy? What is your target market's buying power (demographic research)? What are the current spending patterns of your target market? Are your customers buying less or more from your website and why?

Socio-cultural Forces: How are society's (and your targeted market's) demographics and values changing and how will these changes affect your web-based business? What is the general attitude of society regarding the Internet, your business, and its products/services? What ethical issues should you address?

Legal and Regulatory Forces: What changes in various government regulations (domestic and foreign) are being proposed that would affect the way you operate? What effect will global agreements such as NAFTA and GATT have on your web-based business?

Technological Forces: What impact will changing technology have on your target market, if any? What technological changes will affect the way you operate your website, sell your products/services, and conduct marketing activities?

Identify Target Market: What are the demographics of your target market, i.e., characteristics such as, sex, age, income, occupation, education, ethnic background, family life cycle, etc.? What are the geographic characteristics of this market, i.e. its location, accessibility, climate? What are the psychographics of your niche market, i.e., attitudes, opinions, interests, motives, lifestyles? What are the product-usage characteristics of this market?

Needs Analysis: What are the current needs of your target market? How well is your website and its products/services meeting these needs? How are your competitors' meeting these needs? How are the needs of your niche market expected to change in the near and distant futures?

Market Positioning

Your website and its products/services cannot be all things to all people. Look at margarine or aspirin, for example, and the extremes that have been taken to create brand awareness and product differentiation. Marketing requires continual vigilance. Your marketing position must be able to change to keep up with the current conditions of the market. Constantly monitor what is happening in your "space" so that you always have up-to-date knowledge of your marketplace. After you accumulate accurate information about your customers, the segments they fit into, and the buying motives of those segments, you can select the marketing position that makes the most sense.

Performance Analysis

At this moment, how is your website performing in terms of sales volume, market share, and profitability? How does this compare to other websites in your "space"? What is the overall performance of your entire competitive marketplace? If your website's performance is improving, what actions can you take to ensure that it continues to improve? What are your web-based business' *strengths, weaknesses, opportunities, threats*? (In marketing circles these four terms are commonly lumped together into the acronym — *SWOT*.)

Marketing Objectives

Once you have answered the above questions, you are ready to set out your marketing objectives. Define your current marketing objectives. Are your objectives consistent with recent changes in the marketing environment and/or needs of your target market? What is the specific and measurable outcome and time frame for completing each objective? How does each objective take advantage of a strength or opportunity and/or convert a weakness or threat? How is each objective consistent with your web-based business' goals and mission?

Marketing Strategy

Next comes your marketing strategy. Once you have completed the research on your target market with specifics such as demographics, geographics, psychographics, product-usage characteristics, justifications for the selection of this target market, and your competitors in this market. Next consider your marketing mix (pricing, distribution and promotion strategies). How does this marketing mix give you a competitive advantage in your niche market? Is this competitive advantage sustainable? Why or why not?

Other Elements to Consider

Some other elements that fit within a good marketing plan are:

* Distribution channels.
* Pricing and terms of sale.
* Promotion and advertising plan.
* Marketing budget.
* Inventory selection and management.
* Visual merchandising.
* Customer relations.

When drawing up your marketing plan, think about where you want your business to be in three years and how you plan to get there — that's your marketing plan in a nutshell. Marketing your web-based business is a never-ending task. Once you have your information and your marketing plan in place, you must continuously revisit, revise, refine, and revamp it to accommodate changes in your marketplace.

With a marketing plan in place you have a considered strategy to outmaneuver your competition by capitalizing on their weaknesses and emphasize your web-based business's strengths. By increasing market awareness of the offerings of your website, you acquire new customers.

✪ IMPLEMENTING YOUR MARKETING PLAN

Now it is time to implement your marketing plan. Identify a marketing team. This team can be formed from your business' internal personnel and consultants, and people who have intimate knowledge of the Web, web marketing and (if you are a click-and-mortar) how the Web can be integrated with current marketing plans. In addition, find graphic designers, copywriters, and illustrators. Once the team is in place, it should plan and focus on strategic revenue goals. Don't let the team focus on the number of hits your site receives due to an overall advertising, marketing and PR campaign while ignoring whether the campaign achieved the expected revenue goals. Make it clear as to who has the decision-making authority, who is responsible for what, and set out a clear timetable for each campaign's completion?

Don't forget to coordinate your marketing activities. What does this mean? Here are a few example: If you plan to unveil a campaign during the Super Bowl you must ensure your website and your web-hosting service and/or ISP have the facilities to handle the added traffic. If you are a brick-and-mortar, how do you coordinate special promotions? If you plan to offer a specific item as, let's say, a two-for-one promotion, take steps to assure that there is adequate inventory. Also additional contact center help is mandatory when launching any kind of new campaign.

Marketing Venues

Your marketing plan should include marketing through several venues, such as print media, banner ads, affiliates, television ads, radio ads, newsletters, email, etc. However, with the Web's extensive capabilities available for a reasonable cost, make it the cornerstone.

Consistency

Create a consistent marketing message for your website that reflects its mission and goals. This theme will provide consistency to your presentation and continuity over time. The theme can remain constant, although the look and content will certainly change. Always keep the look and content focused on your potential customers' wants and needs and take advantage of any of your competitions shortcomings.

Analyze

Your marketing plan should become the basis for analysis of whether customer needs are being met through analyzing sales trends, customer's comments, return numbers, requests for out-of-stock merchandise, repeat customers, surveys, etc. At some point consider offering new products — either related or unrelated to current ones — and go after new target markets or penetrate current markets more deeply.

Brick-and-Mortar Advantage

If you have a brick-and-mortar business, make use of it. How? By making certain that your website information is prominent in all of the advertisements for your traditional business and check that your website's address is prominent on all your marketing and advertising material, business cards, letterhead, envelopes, brochures, shopping bags, giveaways such as hats and pens, etc. Also make your employees aware of the importance of promoting the online business.

If you are the owner of a successful brick-and-mortar brand, then you already possess a key advantage in you online venture — name-recognition. Just this fact alone will help make your website more valuable and allow it to obtain profitability more quickly.

Finally, institute a procedure to monitor the success or failure of your marketing activities. Will a formal marketing audit be performed? If so, what will be its scope? Will specialized audits be performed? If so, which marketing functions will be analyzed?

⬦ ON-SITE MARKETING TECHNIQUES

How do you keep your customers coming back for more? On-site marketing of your website involves search engine submissions, strategic links, optimizing your copy, banner ads, opt-in email, client retention services, affiliate programs, press releases, and much more. In other words, it's whatever will work to draw potential customers to your website then converting them into paying customers. You want to build brand awareness, foster relationships with your customers, encourage repeat business and return visits to your website.

Established brick-and-mortars that have moved their business (or elements of their business) online must take care to reflect the essence of the brick-and-mortar brand, work with the brand, and make sure that you take the brand online appropriately. In other words, don't jeopardize your product by using marketing gimmicks online that you would not do use for your traditional business.

To keep your customers coming back for more, include in your marketing plan strategies for:

* Targeted opt-in email, which means that the recipients have specifically requested email relating to a particular topic.
* Online newsletters.
* Content Updates.
* Incentives, contests, and surveys.

Some examples are a newsletter for registered visitors, a free gift for answering a questionnaire or for a referral, or a monthly drawing for one of your products. All of these suggestions (out of many that can be implemented) not only build traffic but also start you well on the road toward collecting data on your customers' demographics and offer the opportunity to amass your customers' email addresses for use in future marketing campaigns.

Solicit Customer Feedback

Aggressively solicit customer feedback. It's a great marketing tactic. Here's a suggestion on how to create a good customer feedback campaign:

Create a short, say 25-question, customer survey that has a prominent position on your website. When customers take time to fill-out the survey, thank them by issuing a $5 gift certificate or some other giveaway. Compile the information obtained from the survey and then use the results to fuel growth and change so that your site is always new and exciting. Surveys also can indicate what your customers like and dislike about your site.

Another benefit from using the survey method is acquiring your customer's information —name, email address and maybe even their snail mail address — all very useful for sending out future promotions. But be careful not to annoy your customers. It is advisable to mitigate the possible irritation caused by direct-mail (either email or snail mail) by including something your customers will appreciate such as a newsletter or a certificate for redemption of a small gift or gift certificate; which can, in turn, result in customer appreciation.

Another way to use your survey form is for testimonial feedback that you can then use as content on your website. For example, if you offer a certain brand of shoes, in one of your surveys you can ask for opinions or comments about that brand. Use those comments on your website, such as, "I didn't need these shoes but they reminded me of when I was a child," or "I just wanted them because they were different."

Links

As mentioned earlier, numerous search engines use link popularity when ranking websites. Thus instituting a strategic linking program is a must for any new website. When a search service sees a website that has a lot of other websites linking to it, it naturally assumes this profusion of links means it is a site with compelling content and is well-regarded in the Web neighborhood.

Begin by negotiating reciprocal links, especially with websites that appear consistently in the "top twenty." Having a good base of incoming links from other websites is as important as providing links to the "top twenty" on your website. Remember this can also include your competitors' sites. Why? If any of your competitors' sites are one the "top twenty" returns when a search is done for "shoes" leading customers to that site and they don't find what they are looking for on your competitor's site, but see your link, guess what — they will end up on your site. You will be surprised how many of your so-called competitors will be very glad to link to your site in return for a link back. Why? Because the next time *your* website may show up in the "top twenty" when a potential customer uses a different search criteria.

Work hard to develop link partnerships with websites that are popular with your customers. How can you know what's popular? Use those surveys!

Quality links — links that provide a valuable resource your customers will appreciate — are the only links you should consider when developing your links strategy. These links give you the opportunity to provide a useful service for a potential customer even if there is no sale made at that specific time. Customers remember the websites that enables them to accomplish their goal — even though they may have made their purchase on another website. But because you took the effort to provide your customers with a valuable service — links to quality websites that target the same niche market — they will come back. Eventually, they will make a purchase from your website.

There are many ways you can exchange links. Links can be banners, links can be placed in emails, links can be placed within informational text on yours and others websites, links can be an award, links can be provided within a buyer's guide or directory.

Okay, now you get the idea, so go for it. Try any variety of these links or all of them — just do them tastefully and don't let the links distract from the presentation of your products/services. Links not only help with search engine ranking, they feed traffic to your site directly, and they help to create name recognition. All good things!

Now, how to go about generating links? First decide:

* Which websites do you want as link partners?
* Which websites would be of value to your customers? For example ask your customers, in a survey,
  - Where do they go for information and resources?
  - Where do they shop in the traditional world?
  - What other websites do they visit when seeking the same product/service?
  - Which websites offer products/services that compliment your offerings?
  - Which websites do they consider to be your competitors?

When vacillating on whether to establish a link relationship with a specific competitor, remember that if you offer a quality product that is competitively priced, a link to a competitor's site won't hurt you. It can actually help by providing your customers the information necessary to make a purchasing decision. Your customers aren't stupid; the Internet is a great venue for comparison-shopping, so customers will probably be visiting your competitors anyway. Make it easy for them; they'll appreciate it. Remember *Miracle on 42nd Street* — Gimbles referring customers to Macy's and vice versa — same theory here.

Before searching for link partners, consider what kind of partner you want and what you have to offer the potential partner. Be diligent, be open to opportunities, and always be on the lookout for a potential quality link relationship. If you are on a website that you think might have potential, act immediately, send an email, use their feedback form.

Banner Ads

Banner ads are just small digital billboards that one website pays (in one way or another) to be placed on another website. When potential customers click on a banner placed within another website, they are sent to your website.

Various studies have shown that the most powerful word in a banner ad is "free," that simplicity sells, and graphics enhance a message. If a banner is designed correctly, it doesn't distract from the message. Other elements you need for a good banner design include:

* An attention-getting element, but not one that is antagonistic.
* A call to action.
* A reason to click through.
* Content that ignites clicks, i.e., it tells them to give the banner a click.
* Placement of your website's logo somewhere within the ad.

When a potential customer clicks on your banner ad to visit your website, it is called a "click-through." The ratio of impressions to click-throughs is called the CTR (click-through ratio). An effective banner ad design and astute placement on a compatible high traffic site contributes to the obtainment of a high CTR, which can double the click-through rate of your ad, thereby doubling your return on investment (ROI). A study by Doubleclick.com found that:

* After the fourth impression, response rates dropped from 2.7% to under 1% (banner burnout).
* You need to focus on four important issues (creativity, targeting, frequency, and content).
* The use of simple animation can increase response rates 25% — just be sure that the animation doesn't slow downloading of the ad.
* The use of cryptic messages can increase CRT by 18%, but probably do not attract potential customers or reinforce branding.
* The use of humor is very effective.
* Using a question within the ad can raise CRT by 16%.
* Using phrases such as "Click Here" tend to improve response by 15%.
* Offering free goods or services generally improves CRT.
* Using bright colors in the design is more effective.
* The use of a message that gives a sense of urgency actually decreases the CRT.

If you aren't careful, banner ads often fade into the digital woodwork — even if you use animation and other special effects. If not managed right, banner advertising can be expensive and ineffective. Be very careful with your ad placement and make certain that you do not overpay for ad placement on websites that do not produce real customers to your site. Place banner ads on websites that you know your target customers visit. Targeting equates to better-qualified customers, or potential customers that are more likely to complete the sale. How can you know — via your online surveys.

Note that a study by ZDNet found that animated ads generated CTR at least 15%

higher than static ads, and in some cases as much as 40% higher. However, animation does not take the place of response-driven copy and a creative idea. Avoid animation that takes a long time to download and offers poor copy. But with simple, creative animation more potential customers notice and pay attention, even if they don't click on it. Also, potential customers are more likely to click for more information if the animation effectively emphasizes what product is about. However, the animated ad must have a strong, well-crafted message and a clear call to action.

Frequency (the number of times a viewer sees an ad) is an important factor when planning a banner campaign to build your brand awareness. In addition to maximizing your ad dollar, controlling ad frequency can open up new creative doors by allowing you to create a custom banner package for your branding campaign. The correct delivery frequency is necessary for banner ad success and will help determine how successful your branding campaign will be. The right frequency is usually crucial for getting your message across since too few impressions and your message just isn't seen by enough potential customers, and with too many your banners begin to fade into the woodwork. This is referred to as "banner burnout," i.e., a banner no longer offers a good ROI. Several companies offer help in managing your banner ad frequency rate. One such company is the previously mentioned DoubleClick.com whose services allow you to control a sequence of banners that can be served to viewers in a specific order.

Banner ads on popular websites (especially those that service the same niche market) can run as low as $249 for 250,000 impressions. An "impression" is the number of times surfers see a page that your banner is situated on. Ad placement services such as BannerCell.com, Insidewire.com, Internet Advertising Solution (IAS) (http://iaswww.com), NarrowcastMedia.com, and ValueClick.com can assist you with specific, targeted placement of your banner ads.

The industry average of CTRs is between 1.5 and 2.5%. If you purchased 50,000 impressions and received a 2% CTR, your website would receive 1000 potential customers. It is possible that with a well-designed and effectively targeted banner ad campaign you could receive a higher CTR, which is why a good banner design is important. When using services such as the ones listed in the previous paragraph, you can easily rotate two or three banners during a campaign, which keeps a single banner from going stale. However, to receive the 1000 potential customers requires a substantial investment on your part. Therefore, for small websites, it would probably be more cost effective to use alternative marketing methods and then supplement that with a few well-placed banner ads.

In a study from another ad placement service, Webreference.com, concerning effective ad placement, it was discovered that the placement of the ad on the side of a web page, next to the right scroll bar, increased click-through an average of 228% over placement at the top of the web page. Of course, as viewers become accustomed to the ad placement along the side of a page, it might lessen its effectiveness. Another study indicates that an ad placed one-third of the way down on a page gets better results than ads placed at the top of a web page.

Don't forget to factor in the costs of an agency to create the banner ad and another agency or consultant to place the banner ad and to plot your online strategy.

An economical approach is to develop reciprocal relationships with other websites that will cost you only your time (plus the cost of the banner itself). If you do pursue this method, follow these three rules:

1. Create a banner ad with the criteria set out above.
2. Partner with websites that complement your site.
3. Keep your expectations low.

Consider using banner exchange services that enable you easily to advertise your website on a variety of other sites for free in exchange for hosting other member's banners on your website. Most of these services operate by membership and then provide the means for you to swap banner ads with other members. Some services to check out are ClickEZ's Banner exchange (www.clickez.com), LinkBuddies.com, TrainXchange Network (www.ntwp.net/trainxchange), and WorldBE.com.

Almost all e-commerce websites can benefit from an effective banner advertising campaign. Here is how:

Stretches your ad budget. Banner ads cost less to create and place than other forms of advertising and they also often deliver a more targeted audience than more expensive advertising venues such as television, print ads, radio spots, and direct mail.

Gets your e-commerce's brand where you want it to be seen. It is easier to build a higher traffic volume by strategically placing banner ads on websites that relate to your product/service offerings. For example, an e-commerce site that offers trendy sunglasses can place ads on fashion sites, or a software vendor can advertise on tech sites such as ZDNet.com and Cnet.com. Online ads give users immediate satisfaction by allowing them to just "click" and they are at your website — instant gratification.

Provides quality traffic. When a person clicks on a banner, they are interested in what the banner has to say, and thus what your website offers. Since banners tend to deliver highly targeted sales leads, even ads with low response rates can be very effective.

Helps to establish your brand. If you use the same theme (look and feel) for your banners that you use for your website, the ad can serve to reinforce your brand. Even if a potential customer doesn't click on the banner, they have been exposed to your message, logo, and image. By consistently applying your business's colors, trademarks, and products in your banner ads, you help your brand's image stick in potential customers' minds.

You can easily monitor the results of a banner ad campaign. Most banner ad enablers have the ability to monitor web user response to a specific ad campaign and most will provide their clients with the ability to check web-user response to a banner campaign. By tracking ad performance, you can determine which ads and ad placements work and which don't. This allows you to quickly tweak your campaigns and media buys to improve response rates. Also after you've instituted a banner ad campaign, you can continually test and chart the performance of the banners from the very first day of placement through use of your own log analysis software.

Affiliate Programs

To help drive quality traffic to your new website, consider setting up an affiliate program — sometimes called referral, associate, or partnership programs. No matter what term is used, these programs are attractive to many small websites that want to offset their costs by leasing space for a percentage of the sales made by sending traffic to other online businesses.

Affiliate programs are established by web-based businesses to drive quality traffic to their website. In exchange for a host site sending a customer to your website, you pay the host site a commission based on the sales you generate from the referred customer. These commissions are all over the place and can range from 1 to 30 percent and upwards.

If you decide to establish an affiliate program you should create a system that can do online tracking, so your host sites can count the sales they have generated plus their commissions. The higher you can make the commission, the more likely your host site will actually promote your website in some way so it can send paying customers your way and earn additional income.

It also can work in reverse. With an affiliate program you can make your website more valuable to your customers by offering them goods and services that are likely to interest them. You can earn additional revenue from your site through partnerships with websites that will compensate you for the traffic, leads, and sales you send them.

You can either develop your own affiliate program or contact a service that manages affiliate programs by giving websites access to their own network of affiliate sites and measuring the sales each affiliate generates.

For those that want to go the managed affiliate program path, here is a small sampling of the services you might find when you use your favorite search engine and the keywords "affiliate management": Commission Junction (www.cj.com), Floppy-Bank.com, MyReferrer.com, Performics.com, and AffiliateNetwork.com.

Some readers will want to run their own affiliate program "in-house" so they can have total control over the way the program is ran, and avoid escrow fees, per-transaction fees and the like. There are two ways to go about this. Both methods enable you to avoid handing over your affiliate program to a third party.

The first method requires that you have the necessary skills to create your own affiliate program, e.g. craft the CGI script, and then place everything on your web server.

The second method to use a program such as the Ultimate Affiliate (www.groundbreak.com), which is a CGI script written in the Perl programming language that allows you to set up a fully featured affiliate program on your own web server. Ultimate Affiliate's CGI scripts are easy to install and intuitive to use for the average web developer. The resulting program lets you run a "pay-per-sale" or "pay-per-lead" program where you recruit affiliates to promote your website, then pay them a percentage (or a flat fee) for every sale or lead generated by their referrals. The Ultimate Affiliate program tracks referrals by both cookie and optional IP tracking - making it virtually foolproof for an affiliate to lose a commission. And for the hard-core developers, the script will also create replicated websites for your affiliates to ensure 100% guaranteed referral sales.

Newsletters

Newsletters (whether called eNewsletters, ezines, email newsletters or electronic newsletters) are an effective way to build goodwill and keep in touch with potential customers. Of course, you must provide a form for customers' opt-in (ask for you to email them the free newsletter on a regular basis). Through the combination of a little public relations, image-building and selling, you can provide your readers (customers and potential customers) with information they will appreciate and find useful. However, information is the key word here.

Through a newsletter (which can also be sent via snail mail) you can inform current and potential customers about different topics while subtly promoting your products/services. Your newsletter must contain a main story, which is informational, and then the secondary stories can target your products and services.

Newsletters, when used as an opt-in/opt-out program, are a great tool for bringing customers to your website, cultivating repeat business and building customer loyalty. A good newsletter is entertaining, while providing valuable, worthwhile information. It must be attractive, well written and relevant. Newsletters that are informative and rich in content can be just the reminder needed to remind customers to revisit your website.

Before taking on the task of writing a newsletter you should ask yourself — do I have the time, aptitude and resources or the staff necessary to write a regularly scheduled newsletter? If the answer is yes, start your research. Find other newsletters that have the style and content that you like; read and study them until you know why each particular newsletter is effective.

Write your newsletter to a target audience. If your website serves a specific niche market, then you have no problems. However, if you have a more generalized website, dig into your customer data to find a specific subject your customers will find of interest, and which provides enough material for a regularly scheduled newsletter. Once you have your target audience in mind and the niche subject matter, don't waiver. Stick to that particular subject matter; don't ramble. If your subject matter is collectible Barbie dolls, don't try to work in collectible Hot Wheels. That is another subject. What you could do is a survey and ask how many readers are also interested in other forms of collectibles, if you get a favorable response then start another newsletter for that niche market.

To establish a readership comfort level, create a distinctive style (both writing and layout) for the newsletter, which should be consistent in each issue. Establish a publication schedule (weekly, monthly, bi-monthly). Every issue should provide links to other sites of interest and/or reviews of sites that might interest your niche market. The text needs to be content rich. For example, you can review a new book and give links to sites with other reviews on the book and sites where you can purchase the book. Remember your readership wants a newsletter that informs, educates or helps them in their daily lives, not an "in your face" product brochure.

Create an easy-to-locate newsletter archive on your website so that your customers can find any back issue they might need. Have links to your archive pages throughout your website. In each new issue of your newsletter refer in some way or the other to a back issue; this will lead readers to your archive pages and conversely to your products/services. Additionally, these back issues can help build traffic to your site if you are diligent in registering each issue with the search engines.

A newsletter is a good tool, especially for the small web-based businesses. Incorporate your newsletter into your overall marketing plan. If you produce a high quality, informative newsletter, and let other websites know about it, they will happily provide a link to your newsletter pages. You can also trade your newsletter for links, especially to other websites serving the same niche market (this will help build readership and, in turn, traffic to your site).

Benefits of publishing a newsletter, other than the obvious ones set out above are:

* Your newsletter is distributed via the one item everyone on the Web has in common - an email address. Email is easy, free, and works without a hitch. Readers can look at it at their leisure, print it out, save it, file it, and refer to it when needed.
* You can send potential customers emails (the newsletter) chock-full of informative content about a subject the receiver has shown an interest in (along with a little promo for your website and its products/services).
* You promote not only your website but also its products/services. It is vital that you *subtly*, but relentlessly, reinforce your website and its products/services brand(s).
* You build trust since familiarity fuels the comfort index. Every time your readers receive your newsletter wherein you have shared valuable information, you build and reinforce their trust.
* A newsletter can become an additional source of revenue. You can sell ads in your newsletter, but you first must build your readership to at least 5000 to make ads attractive to other businesses.

You will find that your customers and readers change their email addresses more often than their physical address. You have to find a way to keep track of everyone's current email address. One way is to use a listserv program like Majordomo. Also let your customers and readers in on the action, give them a simple form to fill out that allows them to subscribe, to unsubscribe, and to change their email address.

Market your newsletter, not only should you place your newsletter section with search engines but you also should list it in every applicable newsletter and/or ezine database available. For example, Pertinent.com maintains an ezine (newsletter) directory that serves as a centralized source of information for ezines online. You can find specific information within the directory or promote your newsletter by adding your listing. You also can also use the directory to trade advertising with other ezine publishers in order to increase your subscription base. Another site you might want to check out is ezinedatabase.com. Although the website is poorly designed, the database might be better designed. Finally, look at A.U. Publishing's website's ezine database (www.aupublishing.com/ezinedb.shtml). Registering your newsletters with such websites may help to bring in more subscribers, i.e., more customers.

If you continually publish a quality product, your readers will repay you through purchases on your website. Finally, ask your customers for feedback and always include an opt-out link in each newsletter.

List Management Software

Sending out your newsletter can be a daunting task unless you take advantage of email list manager software. Email lists disseminate a single message simultaneously to a group of people. With email lists, you can quickly and cost-effectively deliver thousands, even millions, of newsletter messages simultaneously over the Internet. Furthermore, through database integration, messages can be personalized according to each customer's demographic information and preferences. Email list management software automate tasks such as list creation, subscriptions, un-subscriptions, bounce handling of email delivery errors, and the like. Look into email list management software solutions such as:

LISTSERV (www.lsoft.com). The first email list manager, and remains one of the dominant systems in use today. It lets you create, manage, and control your electronic mailing lists; maintain interactive email discussion groups for customer service; provide technical assistance; forums about any topic of interest; and distribute personalized direct email campaigns with targeted information that the recipients have specifically requested.

Majordomo (www.greatcircle.com/majordomo/). This freeware is distributed by great Circle Associates. However, Majordomo is not for the fainthearted. The website does not provide technical support for the software, although it does provide links to user support groups. For further help with Majordomo, check out "Majordomo Newsletters for the Novice," which can be found at www.wilsonweb.com/articles/majordomo.htm. (Bookmark this website as it holds a wealth of information for the e-commerce novice.)

Majordomo automates the management of electronic mailing lists through commands sent to Majordomo via electronic mail to handle all aspects of list maintenance. Once a list is set up, virtually all operations are performed remotely, requiring no intervention upon the postmaster of the list site. Majordomo controls a list of addresses for some mail transport system (like sendmail or smail) to handle. Majordomo itself performs no mail delivery (though it has scripts to format and archive messages).

Petidomo (www.ira.uka.de/~patrick/petidomo/). This product offers one of the fastest mailing list server solutions available along with extensive technical options. Petidomo delivers all email simultaneously rather than one-by-one.

Some of these products (e.g. LISTSERV) also allow you to set up a discussion board or an email discussion group on your website. This enhances your newsletter offering by allowing your customers to communicate with one another.

Email

Marketing, promotional materials, and methodology can be helpful in building a good customer base for your website, but be careful to not burn up your communication pathways. *If used properly*, email is one of the best ways for a web-based business to acquire new customers and grow and build long-lasting relationships with current customers. Start with monthly emails then slowly increase the frequency and specialization of these emails by responding to your customer's requests — because you, as a responsible website operator, remembered to ask for feedback. Although the spam crisis continues to affect consumer behavior online, it does not necessarily cloud consumer receptiveness to legitimate marketers: an overwhelming majority of online consumers receive offers by email and have made a purchase online or offline as a result.

It is possible to set up a list quickly, by hand, if it is small or you can use specialized software (as discussed later). Surprisingly good response rates to target email campaigns are achieved because:

* Email hyperlinks allow recipients to go straight to your website.
* The recipient has proactively requested the information and, therefore, is interested in topic.
* The recipient can opt-out if no longer interested in receiving specific email.

With the right email tools, you can have a candid one-on-one audience to get customers to respond to an offer targeted to them. Check out iMailer (www.emailtools.co.uk), Mail Bomber (www.softheap.com) and, DoubleClick's (www.doubleclick.com) Dart-Mail. While on the DoubleClick website, check out the latest DoubleClick Consumer Email Survey.

One of the great values of a simple email offer is that it has many of the advantages of the over-hyped push technology minus all of its complications. To effectively use an email campaign you must request an email address with each order, each new account, and each inquiry your website receives. However, *be sure to ask* your customers if they would like to receive email messages with special sales information and news about any new offerings before bombarding them with email offers. If you forego this step, customers may feel that you are "spamming" them, and opt to go elsewhere, e.g. your competitors.

EMAIL "NETIQUETTE"

If you are going to use email campaigns in your marketing efforts, it is imperative that you provide the email recipient with a way to easily opt-out of the email listing. As an additional "trust" incentive, state clearly that you not sell your customer list to anyone without specific permission from the customer (although it is perfectly legal to sell your mailing list to others). However, if you feel it is important for you to have the option of selling a customer list you can provide a clickable button giving approval for the customer's information to be sold by you, but don't expect that button ever to be used. Use an email campaign to increase your bottom line, not to alienate potential customers.

Give your customers the option of receiving their messages from you in the form of HTML-enhanced mail since this type of mail can deliver a more persuasive message than a plain-text email. All of the newer email clients can accept HTML email, complete with graphics and layout, so use this technology wherever possible.

✪ DATA MINING

If your budget is generous, you can leverage the information you already have to better predict your customers' behaviors and to track business trends. Data mining discovers patterns and relationships hidden in your data that allows you to see emerging trends and patterns and to help you to develop better relationships by leveraging this knowledge. Your ability to discover and analyze your data will give you the knowledge to grow your business with useful and meaningful information. It's almost guaranteed that at least some of your competitors are using data mining techniques to locate high-value customers in order to increase sales and minimize fraud.

Businesses sit with enormous databases holding various customer data (scattered all over the enterprise). These databases contain, not only a customer's name, address, and email, but also data on their buying trends (what they're currently buying), their surfing habits, the technology they are using, etc. The key to success is unlocking and applying that valuable information. With the right tools you can send quick inquiries into the vast array of information found in your traffic logs, advertising reports, shopping cart, etc., and perform real-time analysis of that data, allowing you instantly to capture information about a customer's behavior. Using this technology, scouting through your various data sources, amassing the data, and then turning it into actionable information, allows you to use the data to deliver value-added marketing. However, mining the data contained in your business's databanks isn't for the fainthearted. It is expensive to implement and for staff to use the data mining tools requires extensive training.

However, when properly used, data mining helps you to gain insight into the (evolving) requirements and needs of your customers through analyses of customer data throughout the entire customer life cycle: from identifying prospective customers to extending and maintaining customer relationships. Then use it for developing and executing personalized, customer-centric marketing programs — programs that truly optimize customer relationships and deliver the highest possible return on investment.

For example, your marketing people might want to know the number of customers who saw the new widget banner ad on your website's FAQ page exactly three times before they clicked on it. Maybe it would be useful to find out how many times a customer will search for something before they decide that it can't be found on your website. If your website requires registration somewhere within the purchasing process or uses technology to differentiate anonymous potential customers, the likelihood of using mined data increases exponentially.

Establishing a data mining program is time-consuming, expensive, and requires the amassing of clean data (from log files, email and ad servers, and customer databases) before any analysis can begin. This can be done in-house or outsourced depending on your needs and budget. Next, since data mining builds models to analyze and predict behavior of customers by focusing on associations and patterns hidden in data, you need a staff that is well trained in data mining techniques.

BestProcesses.com provides an example of how data mining saved an e-commerce company money and helped it to understand its customers. The e-commerce business is a publisher of Internet reference books. It experienced a 30 percent loss in its customer base, causing the company to contemplate several courses of action, including converting its books into online databases at a cost of more than $30,000. After completing a data mining project, the company discovered the problem was not as dramatic as originally thought. The substantial decrease was actually an error in tracking customers. Data mining saved the company from incurring a needless expense.

By comprehensively analyzing data, e-commerce operators can see a bigger picture of what their visitors do and where they go. You can use that information to direct visitors to specific pages within your websites and to present them with personalized content. This helps to increase page views, lengthen visits, and increase sales.

Personalization

Personalization (one-to-one marketing) has the potential to completely revolutionize how a web-based business markets its product to customers and maintains its customer relationships. What exactly is personalization? It is a compilation of detailed behavioral knowledge of individuals and/or groups of individuals with certain like-behavioral characteristics and the use of that knowledge to personalize a customer's online experience. To put it another way, personalization is the management of the customer relationship on an individual customer basis, but carried out in a mass production sort of way. With the help of personalization tools, you can entice your customers to stay on your website, make purchases, click on ads, etc. This, in turn, can lead to more sales, larger sales, more frequently returning customers. Customer benefits include easier access to products they care about — ease of shopping is something customers remember, so they will come back.

For the most part, personalization is more than a MyWidgetPage offering. Instead, when properly used the personalization program is tied into a data mining program, which enables a website to serve individual customers personalized ads and perhaps a re-ordering of a web page's content (which is accomplished in real time) to match that customer's behavior and shopping patterns.

To implement an effective personalization effort on your website you must establish clear goals to point the way toward what to personalize. It takes careful planning. You must be able to predict the wants and needs of the individual customer and target what you want to accomplish with the implementation of personalization on your site.

Do you want to increase customer loyalty? If so, add personalization in such a way as to influence repeat traffic. Are your website's products/services something that customers usually feel that they need to research and evaluate prior to purchase (a refrigerator, automobile, etc.)? If so, personalize any portions of your site that can help in the customer's decision-making process. Of course, to do this properly you need to determine how your customers gather information before making a purchase.

Know your customers: Some will know what they want when they first click on your website. Others should be gently led into an information-gathering process that can aid them in their comparison shopping.

Customer Categories. Your personalization tools need to be able to trace the paths and gather information regarding individual customers who visit your website. In general, customer behavior breaks down into five categories:
1. The impulse buyer (no research).
2. The brand conscious customer (little research).
3. The customers wanting the best value for the money (extensive research).
4. The price conscious customer (extensive price comparison).
5. The window shoppers.

Be prepared to engage the customers in each of these categories so that they will stay on your website and eventually make a purchase. To do this, lay out a different motif for each customer category.

Privacy Issues. Personalization tools can be abused. For that reason, personalization has a "bad rep." These tools must be used with care, or you may find that you have alienated a serious percentage of your market share. Do not secretly profile the customer, do not send spam, and do not trade in personal information that you do not have the right to collect.

Personalization tools can provide much good — when used responsibly. For example, they can be used to create a unique session focused entirely on an individual customer's needs, employing techniques to cross-sell, up-sell, and perform goal-driven configuration on a one-to-one basis. Thus you can present your products/services to individual customers in terms that are almost too good to refuse.

Use personalization to understand your customers' interests, and provide them with solid unbiased information on those interests (without charge or obligation). If your customers like what you are delivering, you have a valuable commodity — a repeat customer.

If you think data mining and personalization tools are in your e-commerce business's future check out:

* Megaputer.com's PolyAnalyst 4.6. This comprehensive and versatile suite of advanced data mining tools incorporates the latest achievements in automated knowledge discovery to analyze both structured and unstructured data. (The megaputer.com site also has a good data mining 101 section.)

* Insightful.com's StatServer, a web-based decision support system based on S-PLUS, one of the premier tools for data analysis, data mining, and statistical modeling.

* DeepMetrix Mining Visitor Intelligence Service (www.deepmetric.com), this is a hosted service. It might be just the ticket for the less technically inclined e-commerce operator.

* MarketMiner.com offers an automated, informative data mining tool that automatically produces a complete marketing analysis from, for example, customer demographics, sales, or response data from a past direct marketing campaign.

* Unica Corporation's Affinium Product Suite (www.unicacorp.com) is a fully integrated marketing automation suite designed to allow businesses to leverage customer information across multiple data sources and interactive touchpoints.

✪ CONCLUSION

Creating a stunning website is not enough to guarantee e-commerce success. You need to get the right visitors — potential customers —to come to your website.

Many readers will find that running a business on the Web will seem daunting at times because there are so many avenues a potential customer can take before they find your website. But if you understand the basics of marketing, establish a sound marketing plan, and put that plan into place, the Web will appear less daunting.

So get to work, develop targeted marketing strategies, monitor the results, and continuously strive to improve your efforts.

Chapter 17

Customer Service

*Technology: The knack of so arranging the world
that we need not experience it.*

—Max Frisch, Homo Faber

Good customer service equals good customer retention and good word of mouth.

The explosion of e-commerce transactions brings tremendous opportunity to all kinds of businesses. The Web has attracted both entrepreneurs looking for a new business model and established brick-and-mortars seeking an opportunity to grow their revenues and to expand their customer base through this compelling online sales channel.

Yet, all e-commerce businesses need to differentiate themselves from the competition. That presents a daunting challenge to most online enterprises; especially those just entering the e-commerce arena. The smart businesses realize that at least part of the differentiation will come from the quality of service they offer their customers.

The successful e-commerce business will adopt good customer service strategies that allow them to build customer loyalty, fulfill a broader range of customer needs, and increase the effectiveness of their sales and services.

Look at customer service like this: Every contact a customer has with an e-commerce business, its employees, website, help desk, call or contact center, or other business-related services, influences that person's perception of that business — online and offline.

The technologies that the Web brings to the forefront present a unique opportunity to create and nurture a special one-on-one relationship with every customer. This is why providing exceptional customer service is the best marketing tool a website has.

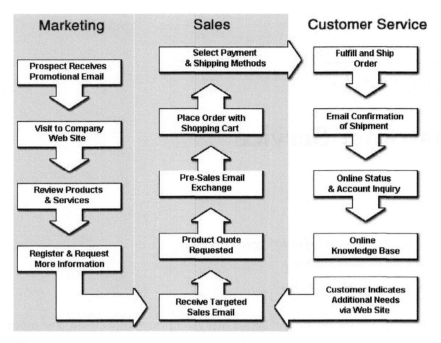

Figure 26. The perfect e-commerce trifecta — marketing, sales, and customer service. An e-commerce business can court customers through promotional material, which entices the customer to visit the website to review online marketing material. Once comfortable with the products and/or e-commerce business, a customer will place an order. If the website is built with the right backend support, the customer can then follow the post-order process through receipt of the product. Thereafter, the customer can re-visit the website anytime he or she has a question, or desires additional information/products/services.

✪ THE PROBLEM IN A NUTSHELL

Online customers bring a whole new set of customer service expectations that never existed with brick-and-mortar stores. When a customer goes online with the purpose of making a purchase, he or she expects to find *easily* all of the information necessary to make an informed purchasing decision. But that customer also expects to hit the order button, provide the necessary information, and boom! The package magically arrives at their door. This is because buying products online feels like magic to many online customers. Trying to manage such unrealistic expectation can provide a customer service challenge.

Let's say that your business is increasing by leaps and bounds. That's good, but the more customers your website attracts means more customer inquiries and more online transactions. That, in turn, translates into a higher burden on your order fulfillment systems and a greater load on your customer service department (which may just be a lone clerk, who not only answers the phone, but also picks and packs the orders).

But success can be your downfall. Despite the pressures put on an overburdened staff of a suddenly successful e-commerce site, you must promptly handle all customer inquires.

Numerous e-commerce businesses are dilatory in their response to online information requests. Once a customer has been disappointed by how slowly their query has been answered, they are unlikely to try again. Almost one-half of the currently operational e-commerce websites provide such poor customer service that they take more than five days to respond to an inquiry or complaint — that is if they provide any response at all! When a website ignores their customers' inquiries, they not only discourage brand loyalty, but actually drive traffic to their online competitors.

Who is your competition? It is the e-commerce sites that respond to their customers' inquiries within 24 hours.

Some sites leave their customers stranded where they can't find an easy way to ask for more information and/or send an email request. Then there are the many sites where the "Contact Us" link simply launches a generic email query screen — with no information about how soon they can expect a reply and/or where else to look for information. Other sites hide their contact information forcing the customer into a "hide and seek game," or give only a webmaster's email address. Some websites don't even give a telephone number for customers to use if they really need to talk to a human. This is not the way to provide quality customer service.

What an e-commerce operator should ever forget is that it takes only a few keystrokes for an unhappy customer to tell the entire world of the problem encountered while dealing with your website. If you have a bad customer service day, week or month — your business is ruined because by the time you've solved the problem the entire buying public has read about your failures via message boards, forums, chat rooms, and even review sites like eopinion.com.

✪ CREATE A CUSTOMER SERVICE STRATEGY

Good customer relations keep your customers coming back for more. Fostering good customer relations requires a customer service strategy that puts in place policies, services, software, and hardware that make it easy for customers to feel confident when making a purchase online. If an e-commerce business doesn't develop a good customer service strategy it will lose customers — the same way a brick-and-mortar would lose them — by not responding to their needs. Efficient customer service is crucial for survival of any business.

Don't be one of the e-commerce operators who put an email form or contact/email link(s) on their website without any plan for handling the increasing volume of email. Admittedly, a small e-commerce business serving a niche market might be able to get away with having no customer service strategy since it probably depends only on email and the telephone for receiving and logging its customer queries and orders. Most websites, however, require an infrastructure that allows them to serve their customers through a combination of email queries, online ordering, and telephone support.

To provide customer-winning customer service and to reduce your customer service burden, encourage your customers to be self-sufficient. The best way to do this is to have a well thought out customer service strategy that provides your customers with the detailed information they want and need. This include product/service descriptions, frequently asked questions (FAQs), and knowledge bases consisting of, for example, product specifications, articles, technical papers and manuals, white papers, and case studies.

Let's now look at how to build a good customer service strategy.

The first step is to ensure that the customer service department and/or call center knows how to handle every aspect of a customer's transaction. Although email should be used as the main means of communication, telephone support (especially a toll free number) is also important. Your customer service representatives (CSR) should be able to access and manipulate all the information involved in a customer's order, including tracking the status of an order through the fulfillment process until it reaches the customer's doorstep. If possible, the CSR should be able to communicate with the customers via email, the telephone, or online direct-connection software (chat) — all of which are within the budget of most web-based businesses.

The second step is to determine just what the web-based business's continuing customer service strategy should address. Start with the minimum strategy set out in the previous paragraph, but also plan ahead. To do this, answer the following:

* Can customers quickly find answers to their most frequently asked questions on the website?
* Can customers easily check the status of an outstanding customer service issue?
* Does the customer service staff respond to all customer emails within one business day?
* Can the knowledge content on the website be continually updated in a dynamic and automatic fashion based on customer input?

✳ Is the most useful and/or commonly requested information presented to customers first?

✳ Do customers have an easy way to get to the human-based customer service?

✳ Do customers consistently return to the website to find information, and if so, is there any way to determine whether or not they do, and how often?

✳ Can the website generate reports detailing the support activities that have taken place on the site on a week-by-week basis? Do those reports help you to determine the ROI of your online business?

✳ Does the website give customers the option to have updates sent to them automatically by email?

✳ Is your website constantly updating its knowledge base by publishing useful information as it becomes available?

✳ Do customers ever praise your website because they found it especially helpful?

If the answer to any of these questions is a "no," "I don't know," or "maybe," it's time to get to work. Begin by monitoring your customers' requests for information as they come in — many tend to ask the same set of questions. Diligent monitoring of customer service inquiries allows a website to determine where to direct its efforts — enabling much more efficient use of human and infrastructure resources. You also can use the information gathered to redesign web pages to make them more responsive to your customers' needs.

Consider building an online knowledge base of product specifications, articles, technical specifications, white papers, and case studies that can answer many of your customers' immediate inquiries. Supplement that knowledge base with a comprehensive FAQ section. This empowers your customers to answer their own questions without human intervention, which translates into not only satisfied customers but also an eventual increase in your business's profit margin. An added benefit of an online knowledge base is that customers develop the perception that the website has a good grasp of what their questions and problems might be thereby strengthening their overall confidence in the site and its offerings.

Don't go down the road that some websites have taken, e.g. www.microsoft.com and www.netscape.com — too often leaving the customer in "self-help jail." Let product descriptions, FAQs and knowledge bases be the first line of service, but in addition offer easy-to-access email support, chat, toll-free number(s), and other forms of direct communication.

You must constantly look for ways to make your website more responsive to the

HOW TO EARN A REPUTATION FOR GOOD CUSTOMER SERVICE

Here are some tips to help your e-commerce site win customer service kudos:

- Try not to depend too heavily on outside suppliers to ship products to your customers. This gives you better control of your order processing procedures.
- Consider providing a 30-day home trial policy for specific products. This is especially a good policy for websites that sell high-end products.
- Extend your return policy during holiday seasons.
- Institute a 30-day price protection policy that guarantees if a customer finds the exact same product for a lower price within 30 days after purchase, you will give them a refund equaling the difference.
- Prominently display your shipping and return policies (use Amazon.com as an example).
- Display all customer testimonials and, if possible, give them a venue where they can post their comments.
- The needs of your customers should always be foremost in every decision you make concerning your website.

constantly changing needs of your customer base. If you don't develop a comprehensive customer service strategy, you will lose your customers to the competition, reap poorer returns on their investment, and even find yourself spending more money than is necessary on expensive customer service solutions such as conventional call centers.

Two Different Approaches

There are two basic approaches an e-commerce business can take to provide award-winning customer service. The first approach is called the "separate technology method," i.e. the customer support is separate from the website. The second is a "multi-threaded contract strategy approach." The second method builds upon the first approach by adding managed email, dynamic FAQ pages, web-based chat, and more.

The Separate Technology Method

When an e-commerce business is in the start-up stage, separating the website from the call center might very well be a good approach. The separate technology method just requires that the web designer place the proper contact information in a prominent place on the web pages. Then either in-house staff or a call center takes product/service orders, handle technical support, customer service, etc.

As the business grows and prospers, however, there will be an ever-growing burden on the business's customer service resources. Automation solutions can help you to keep apace with growth and still provide quality customer service. Automation solutions include direct web-based transactions such as online ordering, shopping carts, etc., followed with a FAQ section to answer the most common questions about the business's products and/or services, etc.

To keep customer service so that it continually compliments the growth of the business (rather than hindering it), the first approach will only suffice in the short-term. So while many new e-commerce websites begin in some phase of the "first approach" mode, don't let your site linger there too long. When laying out your customer service blueprint, keep in mind that your business will need to include an infrastructure that allows it to adopt a more sophisticated technology as the need arises.

The Multi-threaded Contact Strategy

Taking the second approach means that the website is no longer an island, but part of a multi-threaded contact strategy. To provide good customer service on the Web, companies must open new lines of communication. Email, while universal, must be well managed. FAQ pages and a dynamic knowledge base are easy to set up; and when managed properly and kept current (perhaps with the aid of specialized tools) they can be an asset for a website.

Web chat, whether text- or voice-based, can be especially effective for quick real-time customer queries, since it can be faster than a phone call. Examples include real-time query and response (some with co-browsing features) "Talk to Me" buttons. "Call Me" buttons, which allow customers to schedule a call back by a CSR are also effective. While the real-time solution is best, the "call me" system is more affordable and can be implemented by almost any size e-commerce business.

Many small e-commerce sites and even some mid-sized businesses will find the multi-threaded approach to be too much for their budget. But there are e-commerce sites (large and small) that currently use various degrees of this type of advanced customer service strategy. Here are a couple of examples:

Landsend.com though its trademarked "Lands' End Live," which according to the website states: "As you shop at Landsend.com, you can be in direct contact with a customer service representative by phone (if you have a second line or a direct connection to the Internet) or through an online "text chat." Instant assistance!"

VenueSwimwear.com and REI.com both not only serve an international customer base, but also provide their customers a cornucopia of customer service choices: FAQs, toll-free telephone numbers, email and snail mail contact information and last, but not least, online assistance via chat.

So while the full multi-threaded approach may not be within the budget of the start-up e-commerce business, most can implement *some* of the features outlined in this section (and gradually add other features as business dictates) to increase its customer service functionality.

Note: A number of the customer service solution companies and customer relationship management solution companies are moving toward the Application Service Provider (ASP) model, which simplifies integration with a website and thus lessens the initial cost for such solutions.

Live Interaction

Let's look at some of the more common multi-thread solutions.

A "Call Me" button on a website means that call center agents need to be on hand to call back a customer, either instantly or at a scheduled date/time/number. Customers seem to love this feature and it is a really powerful technique especially for a website that sells high value products.

A "Talk to Me" button on a website requires that a CSR be instantly available when the customer clicks the button (the customer also must have a multi-media capable computer — sound card, speakers, microphone and the right software). This option uses voice-over-IP (VoIP) to establish an instant voice link between the customer and the call center agent. However, Internet congestion, delays in the website's servers, in the customer's ISP servers, in the call center's servers, etc. can mean that this method sometimes produces poor voice quality. However, vendors such as Aspect (www.aspect.com), have products that can link VoIP with web chat so if the voice connection goes bad then text chat can take over.

Co-browsing requires that, when requested, a call center agent instantly be available to take over a customer's browser to help with a form or to manipulate text or images on web pages as the customer is viewing the same page. Only businesses offering very high value products (e.g. stock market investments), or perhaps businesses teaching customers to use complex products or who provide online schooling should look into this type of technology.

Web chat is a real time interaction between the customer and a CSR using text to conduct chat sessions. And while this service also requires that a CSR be available basically 24x7, a single CSR can handle multiple chat sessions, perhaps as many as eight. This is good. And since it is highly productive, text-based chat can be used cost-effectively for low end products.

There are many live web-chat solutions available to the average website operator. For instance, Peoplesupport (www.peoplesupport.com), Facetime (www.facetime.net),

InstantService (www.instantservice.com), Liveperson (www.liveperson.com), and others provide live, human customer service agents that are available instantly when a customer clicks on a "chat" button. Most of these services are relatively inexpensive, mainly because most offer a hosted version of their product, i.e. the company hosts the software and then rents it out to you on a monthly basis or other contractual terms, rather than selling the entire package to you for your implementation and management. For example, LivePerson's setup fee is around $1000 plus about $250/month per CSR.

For many e-commerce businesses, a hosted solution is the way to go. Just think, a completely integrated front-end and back-end where the applications are implemented and hosted by a single vendor — that's a model that's hard to ignore. For a small- to medium-sized e-commerce site the ASP route, like that provided by LivePerson, is a good option since all that is needed are a few simple lines of HTML code and the button icon for its web pages, and it's ready to go.

One caveat, though; do not use (at least not at this time) automated online personalities (robots) that supposedly can respond to a customer's live chat questions. They can't — they just cause your customers a lot of frustration. I would advise staying away from this type of solution, at least until the technology improves.

A website could offer its customers that have multi-media capability (sound card, speakers, microphone and the right software) the ability to click on a website button to call a CSR over the Internet (a "Talk to Me" button). Once connected, the CSR cannot only talk to a customer, but if the CSR's computer is equipped with the right software, he/she can even access the same web page the customer is currently on. They can then browse the site simultaneously, solving order problems and even filling out forms together. Customers like the personalized experience (the human touch) this type of technology offers.

Another plus for the "Talk to Me" method is cost savings. The calls are routed across the Internet (voice-over-IP) so they cost much less than a traditional telephone call. So, using the "Talk to Me" option (which is Voice over IP) also will have a positive influence on a website's monthly toll-free number costs.

Note: Numerous industry studies show that shopping cart abandonment rates dramatically drop when some type of "live" interaction is offered.

✪ AUTOMATION IS THE KEY

Everything needed to create a truly responsive site — monitoring and analyzing customer queries, creating the right content and posting it in a well-organized manner, handling email communications, etc. — can be extremely labor-intensive. A website can be a victim of its own success if the volume of communications exceeds the resources dedicated to supporting that communication.

Effective, scalable automation tools can automate such time-consuming knowledge collection, which, when neglected, can result in out-of-date content, and dissatisfied customers, and a lower profit margin.

Automated customer service solutions can be software-driven, services-driven, or a combination of both. They run the gamut from FAQ manager programs, to email management systems that help you automate the process of sorting and responding to incoming inquiries, to full-blown customer relationship management systems. The incorporation of some or all of the systems discussed in this section will help to prevent the disorganization that often arises in many web-based customer service operations — especially problems generated by the inability to handle the volume of traffic that can be experienced if your website suddenly becomes popular.

Most of the solutions discussed herein are expensive. Still you should give them some thought before you decide whether or not to forego the golden opportunity these solutions provide — an online, customer-initiated, one-on-one relationship.

FAQs

Although FAQs are built simply by creating web pages with questions and answers, the knowledge contained in an ever-growing FAQ section will, over time, become difficult for the customer to search and for you to maintain. Therefore, as your website grows, consider automating its FAQ section to ensure that the knowledge contained therein can be searched rapidly by your customers. An added bonus of an automated system is that it allows you to more easily maintain a large and/or dynamic FAQ section.

Software, such as, FAQ Generator (www.cgi-world.com/faq_gen.html), FAQ Manager (www.interlogy.com//products/content/faqman/), and MyCGIScripts' FAQ Manager (www.mycgiscripts.com/faq-manager.html) will provide the assistance you need. For more FAQ management tools go to your favorite search engine and type in "FAQ Manager." The results should provide you with a selection of products that fit your needs and budget.

Email

Customers not only want the ability to place orders for product and services, they also want to obtain additional product/service information, to resolve billing issues, to track down shipments, and to ask installation and product support questions. As stated previously, most of today's websites give their customers information on how to contact them by telephone, email, or by a form on their site. However, as customers become more "Internet Savvy," and as hold times at call centers increase (as they must with the e-commerce explosion), customers will become more inclined to send emails to the web-based business. After all, why should they wait "on hold" to get information (or place an order) when they can just fill out a form or send an email and then read the reply at leisure? Moreover, servicing customers through online methods is a proven cost-effective solution that can give a boost to your business's profit margin.

Most home-based or small web-based businesses can, initially, rely on their internal email system to process customer queries. But even when using this method, the following is suggested as best practice when responding to customer's email inquiries (you should pick and choose what would be suitable for your particular e-commerce business):

* Issue a "thank you" auto-acknowledgement for each email as soon as it is received with a time frame of when an answer can be expected (then respond within that time frame), along with an 800 number that the customer can call if they need an immediate solution.

* Have "click and drag" paragraphs available for CSRs to paste into the email response.

* In some circumstances suggest a phone call (if the customer is willing) rather than being drawn into a long series of email exchanges.

* Above all, be fast — make it your goal to satisfy 90% of your customers' email inquiries inside two working hours, and want to be faster.

Don't make a customer wait days for a response to their email inquiry, or give an automated response that doesn't come close to addressing their needs. Don't require your customer to take an aggressive stance just to obtain proper customer service. While a persistent customer might be able to obtain the answer they're looking for, it will come with a lot of frustration, which doesn't foster long term customer relationships.

Many websites handle their email traffic poorly. Should people who communicate with a company via electronic means receive less prompt service than those that use telephones? Remember, demographic studies indicate that people who browse the Web are usually more affluent than the general public, and are generally inclined to try new things. These are the very customers a company wants to attract and keep!

Note: Some web-based businesses believe (wrongly) that automatic email response can handle all of their customer email inquiries. While automatic email response does have a role in many multi-threaded customer service strategies — it can't do it all — also use live CSR support.

Email Management Systems

Any business that has a large volume of email, or finds that its email volume is growing exponentially, should consider using an email management system. And for those astute e-commerce operators that want to handle email inquiries with the same quality of service as telephone calls, an email management system (EMS) may be in their future. Although costly, an EMS can provide the infrastructure, processes and methodologies to handle online customer service issues in a manner comparable to, and sometimes as good or better than, what is available through most call centers.

As discussed previously, e-commerce is conducted in essentially one of two ways:

1. A customer obtains information on products and services through a website. (How extensive the available online information is can vary from product description to FAQ to an extensive, dynamic knowledge base.) The customer then uses the telephone to call a CSR to place an order and/or to ask for additional information.

OR

2. A customer not only obtains detailed information on products and services through an easy navigational path through a website, but also can obtain additional information through various channels (chat, message boards, FAQs, knowledge bases, VoIP, co-browsing) and, of course, place an order — all entirely through the e-commerce website.

Many customers (especially the Internet savvy ones) prefer to take the self-help route (product/service descriptions, knowledge base, FAQs, message boards, etc.) then, if that doesn't work, they will contact the website via email. If the customer has chosen to make contact via email, then a website must make it as easy as possible by providing (at the minimum) a form on the website for its customers to fill out. A "Contact Us" web page devoted to email address links segregated by departments would be even better. This desire by customers to use email (and online ordering) represents a great opportunity for websites that have implemented a high-quality email management system. However, if implementing an EMS, study the websites of Net2Phone, Microsoft, HP, and AOL first. Learn from their *mistakes*.

The advantage of email (and online ordering) is that the transaction can be conducted in an asynchronous manner. Unlike a telephone call, the parties involved in the transaction do not have to be in the transaction at the same time. This is especially valuable to both parties when the interaction needed is minimal, and the need to complete the transaction is not immediate. For example, customers may want to get the tracking number of an order so that they can check the status of their order at their convenience. An email management system can provide the infrastructure necessary to enable a busy e-commerce website to promptly and efficiently handle the vast amounts of customer-based electronic communications.

Clearly, one of the benefits of having an online presence is that customers can conduct their own research on products and services, at their own pace. As discussed throughout this book, a well-designed e-commerce site makes it easy for customers to find what they want, and helps them get answers to most of their questions. Motivated customers will navigate their way through well-designed web pages, and will make purchasing decisions or resolve a service issue on their own. This kind of self-service system offers tremendous leverage to a web-based business that doesn't have the wherewithal to provide 24x7 call center capabilities. The value of a good EMS can be significant, for both the customer and a busy e-commerce site.

The requirements for a high-performing EMS are to get the right answer to the customers, using the right kind and the right amount of resources. This means that the EMS should recognize the kind of inquiry initiated by the customer, determine what resource it will take to process that inquiry, and then apply those resources to get the answer to the customer. Let us examine each of these requirements in turn.

An incoming message must first be recognized by the EMS for what it is — the system must determine if it falls into a predetermined category of communications, such as a product question, an information request, a billing query, a complaint, or a follow-up to an earlier communication.

Email tends to be free form in nature, so the EMS must have the ability to parse the communication, and determine the category from the message itself. The EMS must then determine if the inquiry can be handled in an automated manner. A majority of customer queries fall within a few categories. For example, a company that sells headsets may receive a large number of installation questions, but the questions themselves may fall into just a few categories (e.g. sound card questions, volume requirements, operating system compatibility, etc.) — the company already knows the answers to most of these questions. With access to a comprehensive knowledge base of answers to these frequently asked questions, a good EMS can match a customer's request to the appropriate answer in the knowledge base.

Not all email can be processed automatically. Some messages will require customization and personalization. An EMS must have the ability to identify emails that don't fit into a known category and route them to a knowledgeable person. That person (CSR) can then respond using the knowledge base, one or more phrase banks that contain standard language that reflects the company's style, and their personal knowledge and experience. The EMS should also provide a spell check function to enhance quality prior to forwarding the response back to the customer. Furthermore, the CSR should be able to annotate notes to either the customer's message, or to the customer's profile. Other CSRs can then use these notes to ensure superior customer service. For example, a CSR may use the annotation capability to record why the customer was not charged for shipping costs if the company policy is "no free shipping."

A CSR may not be able to completely handle all customer communications allotted to him or her. For example, the CSR may prepare a response but want to forward it to a co-worker or supervisor for review before sending it. Another customer inquiry may need a response from someone with more information. To facilitate this, the EMS must provide a workflow capability that allows a CSR to hand off a message to another CSR, or a supervisor. This is similar to what may happen in a call center.

An EMS must enable supervisors to monitor, in real-time, the state of all messages currently open, just as a call center manager can look at the state of all telephone calls being handled by the call management system. This allows supervisors to manage workloads, personnel needs, training needs, etc. Supervisors also can use this information to identify any potential problem areas quickly. For example, a favorable product review on CNET or the mention of a product by Oprah or Ellen Degeneres on one of their daytime television shows may at first cause a trickle then a torrent of information requests. Just like in a call center, an alert supervisor can observe the initial change in the incoming workload, and make adjustments to assure the best quality service.

The EMS also should provide reporting capabilities so that supervisors can more easily measure and manage the performance of their CSRs. Another feature to look for in a good EMS is the ability to survey customers on the quality of service received, and manage the results of the survey. Moreover, the database used by the EMS should be easily accessible so that custom reports can be developed as necessary.

A big benefit of electronic communication is that each message is recorded on media and therefore the details of every customer interaction are retrievable. (This is not often the case with telephone calls). This means that a customer's entire email history can be presented to the CSR every time a customer sends in a new message. This provides a complete context for the CSR. A good EMS should provide not only the ability to preserve and present context, but also give the company a valuable database of interactions that can be mined for marketing purposes in the future. For example, the database of all customer interactions may reveal that customers who ask questions about one particular product often start by asking about a certain feature. This may indicate a need to change the product in some way, or it may even present a revenue opportunity for a complementary product.

Finally, a good EMS should be easy to use, to install, and to manage. It should not require a significant training effort. It should be adaptable and flexible enough to fit into the way you do business, i.e. not require you to change your business models or processes. Lastly, the system should leverage the website's existing technology infrastructure by seamlessly integrating with your current email, e-commerce, and front-end and back-office systems.

While out of the reach for most small e-commerce operations, an investment in a robust, scalable EMS provides returns in many ways for a large or enterprise e-commerce business. An EMS can reduce the costs associated with handling customer service requests by up to a factor of five. An EMS might even allow a website to avoid building or contracting with an expensive call center provider because the CSRs can easily telecommute. An EMS enables a website to assure a consistent quality of service to its customers, by ensuring that all customer interactions use the same base of knowledge and communications styles. It can increase revenues by including new product information with outgoing messages (the Marketing Department will love that feature). Most importantly, it can improve customer satisfaction and loyalty by giving your customers a reliable way of communicating with you.

Is an EMS in your future? You consider investing in an email management system if your website:

* Handles more than 1000 email requests and online orders each day.
* Has more than five full-time employees handling online communications.
* Has a customer-base that expects 24x7 service and the same (or better) level of service as they get from a call center.
* Receives many "routine" email requests.
* Has found that its online communications are out of control.

If you are considering an EMS, begin by looking into Edify (www.edify.com), eGain Mail (www.egain.com), Notions Systems (www.notionssys.com), and Transform Response (www.transres.com/response). There are many more, but this should give you a starting point.

Automated Customer Service Systems

The unique combination of quality automated customer service systems (ACSS) and human customer service teams can help an e-commerce business ensure profitability. The typical ACSS is a web-enabled suite of software that integrates sales, marketing, call center, help desk, field service, inventory procurement, and quality assurance operations. Although these systems are expensive to purchase, when you do the math you may find that, over time, they actually save you money.

Post-sales follow-up technology makes it easy for online merchants to offer superior customer service. These include FAQs and their answers, knowledge databases, and message boards. With the correct deployment, web-based businesses can streamline their service offerings and costs, evaluate and target their key-customer; now a web-based business is on the road to building lifetime value and retention for the long term.

When looking at the implementation of ACSS, take into account that many applications require specialized views of the customer data — predictive modeling tools, campaign management applications, call center packages, and web applications — all which require different views in order to fulfill their role. The customer information architecture must support rapid delivery of customer data in a wide variety of forms suitable for application-specific requirements.

A properly implemented ACSS can find and identify essential bits of customer data that are located throughout a business's data infrastructure. Once found, that data is then imported into a common knowledge base that is shared by each arm of the ACSS to improve every segment of a business's customer service offerings. An ACSS also must be scalable and flexible enough to operate effectively through any number of customer preferred channels, i.e., web, chat, email, fax, telephone or VoIP. The typical ACSS platform is a web-enabled suite that lets a web-based business integrate its customer service and marketing efforts. Research the products offered. To give you a place

OUTSOURCING

Most web-based businesses use some level of technology to increase their customer service offerings. But, technology can't do it all — timely responses are impossible if the human resources are not available. Throughout this book the author has preached "scalability." Well, scalability not only applies to technology; it is also relevant to staffing needs. As online sales increase it will become necessary to increase your human-based assets, especially in customer service. Not only will these needs grow throughout a business's life cycle, they will also ebb and flow with the seasons (for example, holidays, back-to-school, snow season if a website sells snow blowers or mild weather if it sells boating equipment).

As customer service resources become strained to the limit during peak periods, the ability to quickly and cost-effectively scale to meet these demands is essential. One way to assure scalability is to outsource. But be prepared — do your homework and find the call center or customer service solution provider that will meet the needs of your web-based business. If any integration is necessary, have it well in hand before the necessity arises. A web-based business will want a standby customer service solution to have the same access and timely visibility into customer orders as its ongoing customer service solution.

to start, visit Siebel.com), E-talk.com) and Firepond (formerly Brightware) (www.bright-ware.com) and pick up a copy of the author's book, *The Complete Guide to CRM* (CMP).

✪ CALL CENTERS

Significant portions of real world transactions are conducted over the telephone and many of these customers are routed to call centers filled with CSRs who are equipped to handle a variety of customer interactions. Using the telephone for commerce and associated customer service functions is widely accepted by the public. The brick-and-mortar businesses that have migrated to the Web probably have existing call center relationships, but how will these call centers address the world of the Web? If your current call center doesn't meet the task, what steps should be taken to bring it up to speed? Should you use a hybrid system with some matters handled by the call center and some in-house? What if you are a new business, without a call center outlet? As a start-up grows, most will find that they need the services of a call center.

A call center is a combination of employees (CSRs), hardware, and software — all trained or designed to aid in the efficient processing of large volumes of telephone calls. It is where a high volume of calls are placed or received for the purpose of sales, marketing, customer service, telemarketing, technical support, or other specialized business activity. To take it a step further, a call center is a place for doing business by phone that combines a centralized database with an automatic call distribution system. However, a call center can be even more than that, it is:

* A fundraising and collections organization.
* A help desk, both internal and external.
* An outsourcer (better known as service bureau) that uses its resources to serve a number of companies simultaneously.
* A reservation centers for airlines and hotels.
* A customer service department for catalog retailers and e-commerce sites.
* An e-commerce transaction center that doesn't handle calls so much as automates customer interactions.

A self-service website has its limits. No matter how well designed the website may be, not all customers can or will navigate their way to the correct information. And some will not have the time or inclination to sit through an extended browsing session. A website should offer alternatives to these customers. Give them the option of picking up their telephone and calling a CSR. Although this kind of immediate service adds to your costs, it can be extremely valuable to an e-commerce website that offers products and services that require a lot of research or that has complicated installation requirements.

Call center technology and the solutions it provides is a dense subject and beyond the scope of this book. For additional information on call centers go to Online Customer Care (www.olccinc.com), The Resource Center for Customer Service Professionals (www.the-resource-center.com), and Call Center Magazine (www.callcenter.com). All three are have a ton of information — bookmark them. Other websites that you might find of interest include Callcenternews.com and Callcenterops.com. Also use search engines — there is a lot of information on the Web about call centers and outsourcers. Or perhaps purchase one of the many books that are available on this subject. One of the author's favorite is *Designing the Best Call Center for your Business*, Brendan Read (CMP Books).

Multi-Media and Web-enabled Contact Centers

Some of the most exciting trends in call centers are the multi-media contact centers and/or the web-enabled contact centers. E-commerce businesses increasingly will find that their customers want to conduct business in a variety of ways: text chat, email, VoIP conversation ("Talk to Me"), "Call Me", and the toll free number. If a website offers a variety of options to its customers it needs to be sure that the call center it contracts with has the resources to follow through.

Why would a website want or need a multi-media or web-enabled contact center? The Web appears to be a self-contained entity where a customer can surf the web, get information, and buy products and services — it is a great channel or conduit for all kinds of things — but it is a passive channel. An e-commerce site needs a way to handle customers and potential customers who want a dialog, whether through email, chat, telephone, or VoIP.

A brick-and-mortar has a superlative, multi-functional sales tool — the human being. For web-based business to emulate this same type of service, it needs a web-based call center. You're on the Web, let the Web do what it does best — automate. But at the same time keep the human element — this is where call centers enter the picture.

The fact that call centers offer a personal and immediate response makes it an easy choice for customers — it is also your most expensive one. So make it easy for your customers to find quick access to other customer service channels — FAQs, email, a knowledge base, message boards, chat, etc. before they place that expensive telephone call. A well thought out and designed website can help build a contented customer base and at the same time reduce your annual call center expenditure.

Most of the multi-media center and/or web-based call center applications work on the same principle — the website's customers click on a "Call Me" or "Talk to Me" or chat button placed on the website to communicate directly with a CSR. With any of these three options, the CSR and the customer engage in a real-time two-way chat via voice or text. Of course, as discussed previously, for it to be voice-based the customer needs a multi-media enabled computer. Some of the applications even allow collaborative "whiteboarding" or "co-browsing," i.e., the CSR can browse along with the customer, each viewing the same web page at all times — the CSR, in effect, takes over the customer's browser for a short period of time. In addition, the applications usually can automatically track, store, and intelligently queue emails to appropriate CSRs or route electronic messages to the correct destination, so customers don't have to wait hours or days for a response.

PeopleSupport (www.peoplesupport.com) offers a solution that even the small e-commerce business might want to consider (this company offers a wide selection of options). This online customer service outsourcing company's CSRs do everything from managing email to offering real-time text-chat customer service. PeopleSupport offers many flexible services. For example, a website can contract with PeopleSupport for their CSRs (who answer questions as though they were the website's employees) to provide live chat support to help customers perusing your site, but opt to handle all email inquiries in-house. Or a website might want to use PeopleSupport's technology (its software is transparently accessed from its application service provider) and provide its own in-house CSRs to interact with its customers. PeopleSupport provides

CUSTOMER RELATIONSHIP MANAGEMENT

Some readers may be interested in customer relationship management suites — especially those working with an enterprise-sized e-commerce business. If so, look into products offered by Wintouch (www.wintouch.com), Right Now Technologies (www.rightnowtech.com), and Peoplesoft. Note that the solutions offered by these vendors (and others), while providing numerous benefits, are very expensive. To integrate them into an e-commerce business's systems will dramatically increase the cost of operating the online business, sometimes ranging well over a million dollars.

When investigating the feasibility of one of these suites, remember that a web-based business must not only survive the procurement, installation and testing of the system, but the costs for integrating integrate the system with the web-based business's operations. Once everything is up and running you also have the additional cost of maintaining a customer service center and its staff.

For more information on customer relationship management read another of the author's books entitled *A Practical Guide to CRM* (also published by CMP Books).

with many different scenarios within which it can provide an e-commerce business customer service support.

Go to their website and click on their live demo where you can test their chat services through a chat session with one of their CSRs. Also look into LivePerson (www.liveperson.com) and HumanClick (www.humanclick.com).

✪ CONCLUSION

It seems that while many are conquering the basics of the e-commerce business model, they have not managed to come to grips with the practical procedures for handling customer service issues. This is because many website operations don't treat their customers as the source of current and future profits; instead they look at customer service as just a series of individual transactions. This is not the only way to develop and nurture customer relationships.

The Web's key appeal to the general public is its perceived ability to provide immediate gratification. When customers come to a website they expect to find information or products *immediately*, and then to solve the problem or purchase the product *fast*! Consequently, the web user is sensitive to any delay — it only takes about eight seconds before a customer gives up a quest for a product and/or service. Of course, some will continue their search — only on another website. Others will abandon the search entirely.

E-commerce businesses, therefore, operate under tremendous pressure to anticipate their customers' every possible need. This need is passed on to the web designer, the people who contribute to the development of the website's content, the marketing department, and customer support staff. Since the customer is hypothetically everyone in every geographic region, the range of information that potentially may be requested is staggering. Solution? Prioritize — don't get stuck in limbo because of the enormity of the task — get the most important information and services on your website first, and then add to it over time, as your customers' needs direct.

The same customers, who expect instant action and gratification, demand better customer service from a web-based business than they expect from a brick-and-mortar business. To maintain a high level of customer satisfaction the web-based business must realize that "site experience" is actually more important to an online customer than "product experience," and build the website and customer service solutions around that fact.

Good marketing can be for naught if there is poor customer service.

Chapter 18

Logistics and Order Fulfillment

Better later than never, but better never late.

—Anonymous

Sometimes e-commerce businesses forget that the Internet is just another channel from which business can be conducted. Granted it is unique in some ways, but many of the same rules apply, especially when it comes to logistics and order fulfillment.

One of the last functions in the chain of customer-centric processes is the fulfillment of the customer's order. Ensuring that the customer's order is picked, packed, and shipped accurately is the final step in the typical online shopping process.

As online sales continue to explode three factors put new pressures on a website's order fulfillment systems:

* The expansion of the product line offered to the online customer.
* The necessity of moving a large volume of small packages at breakneck speed.
* The task of meeting ever-growing customer expectations.

Websites that don't address order fulfillment with the same energy as they jumped into the role of online selling will find themselves fighting an uphill battle for customer loyalty. E-commerce business can meet these demands by *immediately* taking the necessary steps to develop a fail-safe fulfillment system that delivers end-to-end logistics, i.e., package visibility and service continuity from the "buy" button to the final destination.

End-to-end logistics consist of three basic tenets:

* Empower your customers by keeping them informed and by providing self-service solutions that apply to common fulfillment issues.

* Stay focused on the quality and availability of the product.
* Speedy delivery of the product to the customer.

Running a sophisticated e-commerce site often means operating a warehouse (of course the warehouse may be your garage), pulling inventory from shelves and packing it for shipment, retaining a delivery service, helping customers track an order until it arrives, and dealing with returns. To do this successfully many e-commerce businesses will need to rely to some extent on automated solutions.

Customers have been led to believe through marketing and media hype that the Web is all about convenience. That is why so many come to the Web with the expectation that their online shopping will be quick and easy. Sadly, that is not always the reality.

Whether you are a small start-up business or operating an enterprise e-commerce site, coordinating and delivering at web speed is a tremendous challenge. Trying to meet your customers' inflated expectations makes it even more difficult. But any e-commerce business that succeeds in meeting the challenge gains a tremendous competitive advantage.

✪ IT'S STILL ALL ABOUT THE CUSTOMER

Hopefully, the reader remembers from the previous chapter that customer service is the best marketing tool at your disposal. Logistics and fulfillment is an arm of customer service. Thus to guarantee the best customer service, don't separate your warehouse operations from your order management or from your customer service. This is true no matter what the size of the e-commerce operation.

A small entrepreneurial startup may have little experience with shipping products. A click-and-mortar has an infrastructure in place that is, in all probability, geared toward shipping palettes or cases to distributors or stores — not quantities of ones and twos. Websites have customers with high expectations, they want their products fast, and they may expect shipment of different products to different addresses, and ask for instant credit for returns.

Order Processing and Fulfillment

To be a successful e-commerce business you must be able to process customer orders efficiently and correctly and to deliver the product as promised. This can get messy unless you have the right tools to help you; but many times the e-commerce solution you use to power your site won't provide you with all of the tools needed. To help you handle back-office order processing and fulfillment chores you need tools that can help you keep track of pricing (including special sales), sales tax, shipping costs, items shipped, split orders, back orders, returned orders, drop-shipped orders, customer inquiries, and more.

There are a few shopping cart programs that provide most, if not all, of those functions (as discussed later). But many of you will make do using separate applications — tabulating data from not only your shopping cart, but also using your accounting program paired with a spreadsheet and/or database program. Others will look for some type of application that spans their shopping cart as well as accounting or shipping applications.

First, let's look at two applications that are designed specifically for online order management.

Dydacomp Mail Order Manager (www.dydacomp.com) is design to serve the needs of a small business. It provides a complete suite of accounting modules for mail order and/or e-commerce. As an added bonus, this program also can be used with Yahoo! Store.

StoneEdge Order Manager (www.wilsonweb.com/afd/se_ordermgr.htm) is a full-function order management program based on Microsoft's Access database. If you use the drop-ship fulfillment model, consider this product. The program can be used with a variety of other software including Miva, ShopSite, Americart, Able Commerce, eBay, Half.com, Yahoo! Store, SmartCart, and more.

Now let's consider other order management methods that are popular with the e-commerce crowd.

The first is QuickBooks. Check to see if QuickBooks format order export is available with your shopping cart. Watch the wording carefully because the exact QuickBooks format must be built into the shopping cart system, don't try it if you are told that it is "possible," unless you are an experienced programmer (or you are willing to hire one for the job).

The next option for online order management is the use of spreadsheets and databases. Before deciding to use this method, you should be aware that it usually requires re-entry of data, which not only uses up valuable time, but also introduces another venue for order errors.

Some of you will want more sophisticated order processing and fulfillment features. That means a product that contains integrated shopping cart and order management features. If you decide to go this route, you should understand that the level of sophistication is not the same from program to program. Look for a program that allows you to, at minimum, mark orders as "shipped," "back ordered," "drop-shipped,"

THE ORDER CAPTURE PIPELINE — A SCENARIO

Once the customer order has been received, it's dropped into the "order capture pipeline." From there the e-commerce site must send each product order (there may be a number of "product orders" within each customer order) to the correct fulfillment provider, i.e. a fulfillment service provider (FSP), a drop-shipper, and or an in-house fulfillment center. That provider then assumes responsibility for delivering the product(s) to the customer. Even if that sounds simple, it's not; the system must contain business rules to determine which fulfillment provider(s) to send what portion of the customer's order to. An example of rules:

- Use the same fulfillment provider for all the products, if possible.
- Use the fulfillment provider that can give the customer the lowest per unit cost.
- Check in-house inventory before using an outsource solution.

Remember, if more than one fulfillment provider is used, then the customer will receive separate shipments (more shipping expense). Then, sometime during this process (it varies with systems and is dependent upon whether all the fulfillment providers support real-time inventory queries), the website composes an XML message and transmits it via secure HTTP to the appropriate outsourcers. If the website has outsourcers that do not support real-time inventory queries then the site will need to rely on daily supplied inventory profile data.

Now this is where it gets sticky — and where sometimes there is a need for consultants and programmers. Each interaction requires mapping data from the e-commerce operation's order management system (and pipeline) into messages that can understood by the fulfillment providers' systems (including in-house systems). First, the data has to be represented — the way that is handled is dependent upon the provider's systems. For example, a specific "flavor" of XML needed to represent the transaction, or a EDI X12 format might be used, or even a pro-

"suspended," "cancelled," etc. The following products can handle some or all of your order management duties. (Test this type of application to see if it meets your needs before laying out your hard-earned money.)

Open Source e-Commerce Solutions (www.oscommerce.com). This product combines open source solutions to provide a free and open e-commerce platform that includes the powerful PHP web scripting language, the stable Apache web server, and the fast MySQL database server. This feature-packed out-of-the-box solution allows you to setup, run, and maintain online stores with minimum effort and with absolutely no costs or license fees involved.

Nexternal Solutions (www.nexternal.com) provides a hosted e-commerce solution. Its shopping cart can be used to sell directly to consumers or to other businesses. The product maximizes order conversion (orders / unique visitors) coupled with tools that increase site traffic. Since it is a hosted solution, the vendor handles the technical infrastructure and provides the back-end management tools.

Flashecom software (www.flashecom.com) is another hosted product that proprietary batch-file data format or a "special" crafted transaction-related variety with its own key-value-pair grammars may be used to represent the necessary transactional information.

The next step in the order process is collection of the customer's shipping address and shipping instructions. Depending on decisions made upstream in the pipeline, the available shipment options are determined by the outsourcer(s). If multiple outsourcers are involved, then a shared shipping carrier is usually used to ship the order. If a single outsourcer is involved, then the website queries it for current delivery options. Other complications at this stage can include support for multiple ship-to addresses for the same order (for example, Christmas, Mother's Day, Graduation season, etc.) as well as gift wrapping and messages (e.g. "Happy Birthday Grandmother, Love James").

Now we are at the payment stage where credit card information is entered or retrieved from the customer database, and the credit card is authorized for the order total. The e-commerce site interacts with a credit-card clearinghouse to validate the credit card and to determine whether there is sufficient credit available. Communicating with a clearinghouse normally requires a dedicated line and transmittal of the requests using the clearinghouse's proprietary protocol.

Once the credit card has been approved and the customer (all of the above should have occurred within seconds) has clicked on the confirmation button, the order is accepted by the website's order management system. At this point, a few options exist depending on the site's policy and outsourcer(s) capabilities. If the site offers the customer the opportunity to cancel the order within a specified time frame, the order is not immediately transmitted to the outsourcer(s). If the outsourcer doesn't support a real-time protocol, then the order will be transmitted during the next batch-file generation cycle. If the outsourcer supports real-time order processing, then the site can transmit over secure HTTP, containing all of the information necessary for an outsourcer to process the order.

vides customizable order invoices, checkout with or without registration, inventory controls, banner manager, affiliate program, web storefront, shopping cart, catalog, integrated shipping system, eCoupons, and much more.

MonsterCommerce Ecommerce Software Shopping Carts for Small Business (www.monstercommerce.com) powers thousands of e-commerce sites. The product comes equipped with hundreds of pre-programmed features to help you manage all aspects of an online business with an easy-to-use, fully customizable web-based interface, including real-time shipping, product editor, inventory editor, payment manager, order status, and much more.

✪ MIRED IN THE SMALL PARCEL DILEMMA

Although online shopping has exploded (recent statistics shows that 2003 should see over $3.2 trillion in Internet sales worldwide from the nearly 75% of Internet users who have shopped on the Net), less than one-half of e-commerce operations make a profit on a one-item product order. Profit comes when the customer purchases more than one item, but even then it is dependent upon efficiencies in the website's logistics and fulfillment processes.

With the right processes in place, an e-commerce website can keep the cost of receiving, storage, and picking down to less than a dollar per product. But the packing cost (labor, the carton, and packing material) and the delivery is another story. Those items run approximately $5.00 to $7.00 per product. A website also has to overcome the cost of continually shipping out small packages where the shipping sometimes is at least one-half of the overall value of the product purchased.

This high distribution cost is one reason why the Web has not progressed past the "rapidly emerging sales channel" stage, except in a few specific areas — books (where 20-50% discounts are given on newly published, popular books), high-ticket luxury items, and virtual products (music, ezines and software).

Unless the customer is shopping for a specific hard-to-find product or finds the convenience of shopping online outweighs additional shipping costs — your e-commerce business will have an uphill battle. For many products, it is much less expensive to go to the local retail store, even when the cost of gas and the time to drive and shop is factored into the equation. An e-commerce business also must overcome the immediate gratification factor — shopping at a local retail store means the product is in-hand — the customer is not forced to wait for its arrival.

✪ LOGISTICS

So while it is easy to build a website that can take secure online orders, getting the product to the customer can be problematic. That last yard — completing the sale to the satisfaction of the customers so they return and order again — takes planning and organization. The key word is "completing" and the sale is not complete until the product is in the customer's hands and the customer is satisfied. It's the least glamorous part of operating a web-based business and it's the hardest to get right. The complex process of getting a product from the website's virtual checkout counter to the customer's doorstep is what order fulfillment is all about.

Logistics moves the right product to the right place at the right time. For an e-commerce business to be profitable, it must reduce its order fulfillment/logistical costs. Don't think you are alone in your struggle; order fulfillment is a logistical challenge for the whole e-commerce community. This is why many of the major brick-and-mortar websites offer only a limited number of products.

E-commerce logistics requires integration of systems that run from your shopping cart clear through to the shipper's tracking system so that you and the customer can

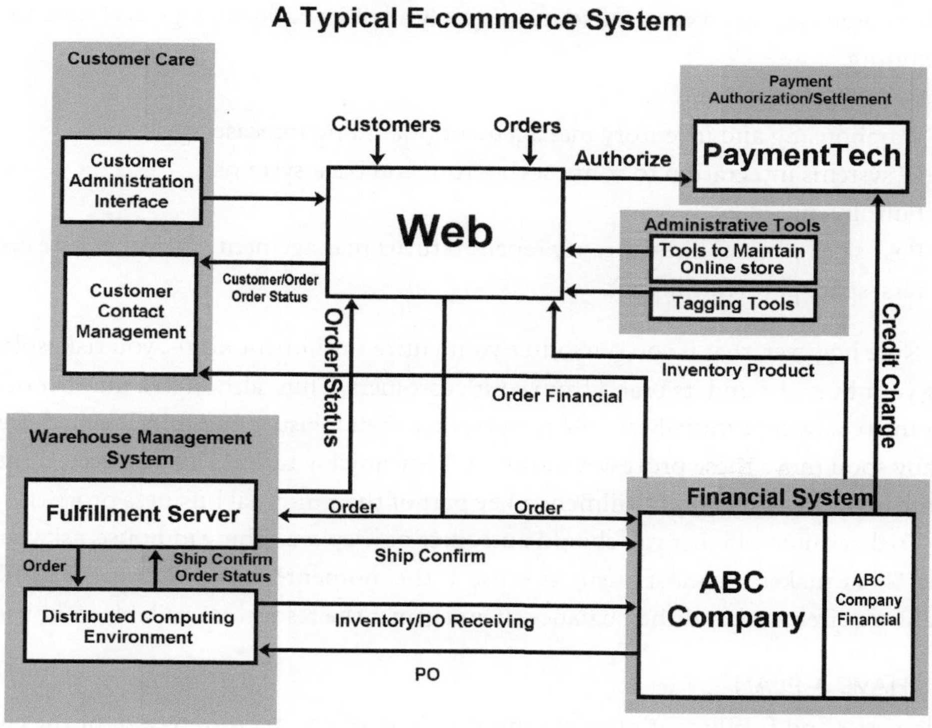

A Typical E-commerce System

Figure 27. This diagram depicts a typical e-commerce system including order fulfillment processes.

keep apprised of a package's whereabouts. But that's just the beginning. E-commerce logistics also requires managing the complexity of delivering daily thousands of mostly small packages to their unique destinations. To do this cost-effectively you must integrate your inventory and distribution systems. But, with few exceptions, e-commerce operators have simply ignored this necessity. Before making the leap into e-commerce you must consider how your business will:

* Process orders.
* Manage multiple suppliers seamlessly.
* Provide a comprehensive exception-handling process.
* Handle the issue of returned products.

Thankfully, an e-commerce operator needn't become an immediate expert in logistics and fulfillment; but someone in the value chain must be an expert and that expert can come from outside your organization. In fact, it isn't that difficult for a new e-commerce business to outsource virtually all of its back-end logistical functions including:

* Order management.
* Warehousing and inventory management (including forecasting).
* IT systems integration to connect the front and back systems.
* Fulfillment.
* Post-sales services like warranty repair, returns management, customer care centers, spare parts fulfillment.

Note however, that if you outsource your entire fulfillment needs, you risk isolating your business and its brand from your customers. Thus, although a number of e-commerce startups initially choose to outsource their logistics and fulfillment chores, many soon move these processes in-house. They quickly realize that to succeed they must make logistics and fulfillment a key part of their internal business process.

To determine whether you should outsource or keep everything in-house, ask yourself: What makes the most business sense at the moment? Assess your order fulfillment requirements and then balance them against the resources available in-house.

✪ HAVE A PLAN

A logistics and fulfillment plan lays out the chain of events from receipt of the customer's order through delivery of the product to the post-sale processes. It should be a guide upon which all fulfillment decisions are based. A formal logistics and fulfillment plan will ensure that everyone considers, not only your e-commerce website's requirements, goals and objectives, but also what each player can bring to the table. This will help to assure that everyone is on the "same page."

You must craft a logistics and fulfillment plan that fits your e-commerce business's specific needs. Thus there are questions that must be answered.

Product and Packaging:

* What are the physical characteristics of the website's products and how do they vary in size, weight, and packaging?
* Does the website offer products that will require special handling and/or packaging (breakables that need repackaging, different size items in one box or a separate box per item, signature on delivery requirement, etc.)?
* Will the website offer products that require a number of different components or accessories to be in the same box (a printer, printer cable and spare ink cartridge or a DSL Modem, a telephone cord, line filters, etc.)?

* Do any of the products require sub-assembly?
* Are any of the products "over sized" or "heavy weight" so that they may require a different carrier from the norm?
* Are any of the products perishable or fragile, requiring a specific transportation mode?
* Do any of the products require special licensing for transport (e.g. alcohol, pharmaceuticals, etc.)?

Average Order:

* What is the size and value of the minimum, maximum, and average order?
* What is the website's current volume of orders?
* What volume of orders is expected 6 months, 1 year, 2 years and 5 years from now?

Shipping:

* What carriers do the website and its suppliers use and what system is in place to track shipments? Is the tracking system easily accessible to everyone, including the customers?
* Will the website offer its customers online tracking and tracing?
* Will the website offer its customers a choice of carriers? Online?
* How will shipping rates be determined? Can the website accommodate customers who already have a specific carrier account number (e.g. UPS, FedEx, AirBorne) and want to be billed directly? Is COD an option?
* Can customers ship to multiple addresses from a single order?
* If the website does ship in branded packaging, will this external branding affect security as far as delivery is concerned (i.e. will it encourage theft)?
* What is the back-up plan if notification is received that a shipment has been delayed/lost/damaged?
* Is the website willing to absorb any of the shipping and handling costs? If so, under what conditions?
* What is the acceptable level of shipping costs compared to the order value (from both the customer's perspective and the website's)?

Delivery Quality Control:

* What is an acceptable time period to get products to the customer?
* What are the criteria for on-time delivery, damaged claims, order accuracy, supply availability?
* How will quality checks be performed?

Inventory:

* How are orders for out-of-stock products handled? If the product is expected to be available shortly, does the website still take the order and risk having to upgrade the shipping to meet the promised delivery date?
* How much inventory is needed to ensure product availability?
* Can suppliers respond adequately to a sudden increase in orders?
* How will replenishment be triggered once an item is picked from storage?

Returns:

* What is the percentage of anticipated product returns?
* How are returns to be handled (dispose of them, use a secondary market, refurbish and put back into inventory, return to supplier. . .)?
* What are the arrangements with suppliers for returns?
* How will customers return the products? Do they require authorization or special packaging/labels for return? Are there any return restrictions, and if so, does your website have the documentation posted to support that policy?
* Who is responsible for inspection/valuation?
* What information systems determine when ownership shifts to the customer or back to the supplier?
* What records are required to keep track of potential tax write-offs or supplier credit?
* How are tracking records integrated with inventory management and shipping records in order to produce necessary documentation?
* What financial arrangements are in place with suppliers for return credit?
* What documentation is required from the receiving department?

International Orders:

* Will your website sell products for international shipment? If so, how will currency transactions be handled?
* Who will fill out international documentation?
* Who will be responsible for duties and taxes/customs clearance?
* Will these costs be included in the shipping cost quotes?
* Have you planned for the additional costs elemental to international shipping/transactions?

Seasonal Considerations:

If the website offers products that are seasonal, how will that affect the warehousing/inventory requirements?

Suppliers:

* Are suppliers willing to ship directly to your customers upon request?
* Can they ship daily, if necessary?
* Are suppliers willing to change from a "pallet" or "bulk" shipment basis to a smaller count shipment? In other words, will the suppliers accept small orders?
* Will the supplier be responsible for transportation costs to the fulfillment center?
* If your website uses drop-shippers, are your suppliers willing to repackage their product in your branded boxes and use your branded labels, promotional material, etc.?
* If your website uses drop-shippers will they insist on using any of their branded products? If so, what and how will it affect the website's brand?

Infrastructure:

* How will your website send order information to the fulfillment service provider, how often (real-time, hourly or daily)? Separately or in batches?
* How will your website handle package tracking and provide the information to your customers?
* How are your website's order entry systems integrated with the fulfillment, inventory management, returns management and shipping systems?
* Do some orders get priority over others, if so, what are the criteria?
* How are your website's financial systems integrated with its inbound shipping, inventory management, orders entry, delivery and returns information systems?
* Can suppliers, order entry folks, inventory management personnel, fulfillment workers, accounting personnel, shipping department employees, and carriers all "talk to each other" to trigger coordinated action?
* What online "alert" systems exist for replenishment, delivery delays, order inaccuracies, erroneous shipping addresses, incoming returns, etc.?
* What system is in place to track all products as they move from supplier to customer (item by item, not just by order)? Options might include bar coding, infrared tags, radio frequency tags. How does the shopping-cart technology link with the warehousing/distribution operation?

Warehousing and Fulfillment Centers:

* What are the warehousing requirements, i.e. the area needed for inventory, supplies and shipping prep, docks, fork lifts, racking and sorting bins, conveyors, pick-and-pack systems, radio transmitted bar coding, software, shipping and labeling equipment, etc.?

* Is the inventory storage area designed for easy and efficient order picking?
* What are the preferred geographic locations of the fulfillment center(s)?
* Is a dedicated facility needed or are the products to be put into a warehouse/distribution operation used by other companies as well as yours?
* Are sophisticated services and space needed or would a direct mail company or public warehousing space be satisfactory?
* How well does the distribution facility accommodate multiple carriers for inbound/outbound shipments? Who will handle scheduling?
* What is the financial trade off between outsourcing your order processing and logistics versus operating your own warehouse and distribution facility?

Insurance:

* What is the website's policy regarding insurance coverage for loss, damage, and theft? What is the carrier's and what is the customer's responsibility?
* Is insurance information clearly posted on your website?

Once the answers are in hand, use them as the starting point in the design of a good logistics and fulfillment system that will efficiently serve you and your customers' needs.

Exceptions

Just as in the traditional sales channel, there is the need for handling exceptions. If your website uses the drop-ship model (explained later in this chapter under "Fulfillment Models"), what do you do when an individual customer order includes products from multiple suppliers and one of the products ordered is out of stock or backordered? Or perhaps you handle your own shipping and inventory and you are out of stock and you don't know when it will be replenished.

How will your website handle a partial shipment? Hold the shipment for all the items, ship all ordered items that are available and cancel the rest? Do you request a customer's input prior to making the decision? Do you hold the order until you receive a reply from the customer?

One solution is to have the order fulfillment system include event-based triggers that are modeled on specific business rules. To implement such a rules-based system requires that you sit down with the appropriate personnel, suppliers and, if necessary, the fulfillment provider, and correlate the business rules.

Management of Suppliers and Channels

Most websites will have multiple suppliers providing a wide array of products. Again, the drop-ship model (although in some instances it is the most efficient) poses the most problems to the web-based business. However, unexpected situations can arise with any fulfillment model, and web-based businesses should have the ability to accommodate customer orders that include products from a variety of suppliers. The website must be able to apportion each order appropriately and to track it through each supplier, and at the same time give each customer a seamless view of their order's shipping data. If a supplier messes up in product delivery, your customers will hold your website responsible, not the suppliers. Thus you must keep on top of your entire supply chain.

Back-end Integration Issues

An e-commerce business must have a clearly defined process for moving an order from its website to its fulfillment center, whether it is in-house or outsourced. This process ties the order to its payment and fulfillment processes — this is where integration with your back-end systems becomes critical. The large and enterprise websites will also find it necessary to tie into back-end applications, such as enterprise resource planning (ERP), inventory management, and financial applications.

A brick-and-mortar that is making its move to the Web may already have pieces of the fulfillment process in place, but the in-house operations may not be familiar with the one- and two-item shipments that will, in all likelihood, become the norm for its web operation. This business group will be tempted to build out their existing "bulk" distribution center to handle the new website's fulfillment needs. While not always a good idea, depending on the legacy infrastructure, this method could provide low ongoing costs with optimal operational control, easy stock status information, and higher product availability. On the downside, there may be a risk of confusion to current operations. Also, the ability for existing personnel to adapt to change should be a consideration. Furthermore, the transition could generate a long period of inefficiencies due to the necessity of building a complex system of individual software components that require integration, customization, and management in order to merge the web-based business with the traditional business's systems.

An alternative solution is to find a fulfillment provider to handle the your website's logistical processes. The decision should be based upon scalability — does the brick-and-mortar (and its website) have the budget and the time necessary to build and maintain its own customized solution? Or can a fulfillment provider more effectively furnish a solution to fit either or both its e-commerce and traditional channel needs?

All web-based businesses will find integration with their back-end systems (content management, customer management, customer service, order fulfillment, inventory management, financial, etc.) to be a costly challenge. Still, such integration is absolutely necessary if you are to avoid fulfillment chaos.

✪ SHIPPING

For the web-based business, selecting the right carrier can be a difficult chore. There are so many options offered that the best choice is not always clear. When looking for a shipping partner ask the shipper if it:

* Offers a full range of shipping services, i.e., same day, next day, ground, air, etc.?
* Has online real-time tracking and tracing that can integrate into your website's back-end and warehousing systems?
* Has an automated and easy return service that is consumer friendly?
* Offers complete transportation management services, including palletized inbound shipments and small package services?
* Provides the same connectivity and tracking services when shipments move outside the shipper's network (a commercial airline or a train or ocean container)?
* Handles orders 24x7? If not, when is the latest the shipper will accept a package or schedule a pick-up?
* How late can a package get into the shippers hands and still be delivered the next day?

It is essential that a website integrate tracking information into its systems and that it makes that data available to its customers. For example, Federal Express has a customized shipping system that allows qualified web-based businesses to track deliveries once every hour. This allows a website not only to check when a package was delivered and who signed for it, but also to track the orders that were returned because of a problem with the delivery itself. This information can be sent to the website's customers via email.

Shipping is one of the most costly factors of order fulfillment. Keep in mind that shipping rates are determined by three major factors: travel mileage, package weight, and delivery times. To better understand the influence of these factors consider this scenario:

Let's say you're shipping a package from San Francisco to New York City and the weight of the package is five pounds or less. Using FedEx Standard Overnight or UPS Next Day Air Saver would cost approximately $30 (FedEx being a few pennies more), Airborne Next Afternoon Service runs approximately $21, and the U.S. Postal Service

Express Mail runs $24 dollars. Second day service is more reasonable, although still expensive — FedEx and UPS both running around $16 (again FedEx is a few pennies more), Airborne runs a little over $10 and the U.S. Postal Service charges $6.50 for its Priority Mail service.

Now let's look at how shippers earn the $6.50 to $30. They:

* Transport the package whether it is 5 miles or 5000 miles.
* Provide tracking information so that you and your customer can know exactly how far the shipping cycle has progressed.
* Deliver the package to the customer's front door.
* Record when the package is delivered to the customer's front door.

Speed is an expensive commodity — but the Internet is all about speed causing many web-based businesses to find themselves in a catch-22 situation — how to arrange a speedy delivery of products, *inexpensively*!

If you offer your customers shipping options, post clear, concise, and fair shipping choices and policies on your website. If overnight service is offered, clarify what time orders must be received for same day shipping and that orders placed on weekends and holidays are shipped the next business day (unless that is not the case).

There is help available. Several online sites provide information on the best shipping options based on origin and destination of the package as well as based on the weight and dimensions. Check out iShip.com. This website offers templates that allow you to find the best options for shipping a package based on delivery times, weight and destination with a comparison of UPS, FedEx, Airborne, and the U.S. Postal Service. Just fill out the online form with the appropriate information — shipment dimensions, postal codes and loss-protection options. It even has a place for a website to add its handling charges (if any). The iShip.com site also offers a tool for online auctions and provides an integration tool that allows e-commerce sites to provide the service directly on their web pages as a convenience for their customers.

A more sophisticated alternative is InterShipper.net. This website helps e-commerce operators and their customers manage all their shipping activities whether domestic or international. A single real-time shipping quote from multiple leading carriers allows buyers to select the shipping cost and service that suits them the best right from their desktops, and enables pre-bid or pre-purchase shipping estimates to expedite purchasing decisions. InterShipper integrates into web malls and stores, auction sites, portals, e-marketplaces, and e-commerce procurement systems, thereby optimizing shipping logistics and tracking processes.

✪ INTERNATIONAL ORDERS

To service international customers, websites need significant expertise in customs clearance, duties and taxes, currency conversion capabilities, and a good relationship with an international shipping company. Thus web-based businesses don't offer international shipping.

For those web-based businesses that want to have a global customer base, the first step is to ensure that their back-end systems are designed for international addresses, international currencies, estimating delivery costs, and so forth. The next step is to tackle shipping. Whatever company you decide to use for your international shipments (and try to make it only one company), that company must provide global shipping to virtually any address anywhere and to help with customs clearance, duties and taxes payment, and with proof of delivery.

Again, there is help available. The previously mentioned Intershipper.com offers help with the choice of international shippers.

If you operate within the U.S. borders, the United States Postal Service's Express Mail International service offers Global Package Link, which establishes a direct link between your e-commerce business and the U.S. Postal Service, enabling the post office to handle most of the documentation for international shipments. This is a worthwhile feature for high-volume shippers. Part of the Global Package Link is the Customs Pre-Advisory System (CPAS), which relieves the website of paperwork and helps speed its packages through customs. CPAS enables customs agents to review the contents of the shipment prior to its arrival and decide if the parcel requires inspection.

For small international shipments, Global Priority Mail is a good choice. The time in transit is anywhere from four to seven days. The U.S. Postal Service offers two sizes of mailers, which are good for magazines, small books, CDs, and any other small, flat item (as long as the package is less than 4 lbs.). Both mailers have a low flat rate from $7.00 to $9.00, depending on the country to which you are shipping.

✪ FULFILLMENT

Although most websites currently receive fewer than 400 orders per day, as the buying public gains trust and comfort with Internet purchases, web sales should increase exponentially. Inventory and fulfillment issues resulting from increased orders will continue to catch many an unwary e-commerce site unprepared. Now is the time to perfect the fulfillment systems and services needed so that when your website's business takes off, everything is in place.

When a customer orders a product from a website, that customer is trying to achieve one or more objectives:

* Save money on product cost (which is partially offset by the addition of shipping costs).
* Avoid the additional expense of state sales tax.
* Purchase specialty products that are not available locally.
* Convenience by avoiding a time-consuming trip to a mall or specialty store.

What your customers fear most about buying from an e-commerce site is the uncertainty of whether their purchase will be delivered when promised. Web newsgroups are full of tales about products ordered and not received when promised (sometimes never).

There comes a time when every web-based business selling a physical product as opposed to a virtual product (such as software or music that can be downloaded over the Internet) must decide how to pick, pack, and ship the product to its customers in a timely manner. As stated many times in this book, online customers have zero tolerance for delays and poor service. But there are steps an e-commerce business can take to provide superlative fulfillment services. They include:

* Linking the website directly, in real-time, to the fulfillment center (in-house or otherwise).
* Building a real-time inventory management system so that the inventory is constantly verified — if a customer sees it on the website, the product should be in stock and instantly available. If it is out of stock, but on order, have a system on the website that can advise the customer that the product is backordered and when the customer can expect it to ship.
* If the customer places the order by 5:00 p.m. (local time), the product should be shipped to the customer that day.
* Orders should be bar coded and scanned during picking, and check weighed at the manifesting scale to insure complete picking accuracy.
* If an order will take more than 2 days to deliver, special low rates on 2nd day shipping should be offered to the customer (i.e., fulfillment center is in Vermont and customer is in Hawaii).
* If an order placed prior to 5:00 p.m. (local time) cannot be shipped the same day, it should be shipped the next day with free upgrade to "next day" delivery to insure it is received within two days.
* As soon as the shipment is manifested, an email should be sent to the customer,

showing shipping date, expected arrival date, final costs and tracking number. If the order is to be shipped from multiple sites, an email message should be sent for each part of the shipment.

* Another email should be sent the day after expected delivery for verification that delivery was made and accurately fulfilled.
* A bar coded "return" label and document should accompany each part of the shipment, just in case any part of the order must be returned.

Outsource what can't be done in-house. Re-visit the situation regularly and when the return on investment (ROI) shows it is feasible, integrate the process in-house with the necessary technology, personnel, and facilities in place.

Keep the Customer Informed

The e-commerce site that understands the importance of the logistics involved in *timely* order processing, *notifying* customers of unavailable items, *providing* real-time order tracking and status information, *implementing* billing procedures and inventory-tracking procedures will soon become a favorite online shopping site. What customers don't *see* on their favorite e-commerce site is its behind-the-scenes rapid and efficient ordering and fulfillment process. But although your customers could care less about the underpinnings of your website and how it works, they do want it to work for them easily and accurately, 24x7.

Order processing and logistics is where an e-commerce business's back-end transaction management takes center stage. Delivering order and shipping confirmations to the customer's email in-box at the time the order is processed should be common practice. Customers should be kept informed as to where their order is in the pick, pack, and ship process.

Order confirmations should contain enough data to give your customers confidence that their order will arrive on time. Once your website has confirmation that the order is on the way to the customer (it is in the hands of the shipper), you may wish to follow with an email to the customer. The email may say something like "Just thought we would let you know your Order Number 12345 is on its way." (Also provide the shipping and tracking information plus a link to use if they want to check on the status of the shipment). When the website obtains confirmation that the package has been delivered, a rigorous customer service will follow-up with another email to the customer confirming that delivery was received and verifying that the product(s) met the customer's expectations.

THE DROP-SHIP MODEL CONUNDRUM

If you use the drop-ship model to provide all or part of your products, you may find it difficult and time consuming to keep up with the "order-processing end." The drop-ship model requires 100% diligence. Your customer does not care that your drop-shipper didn't supply you with current inventory data, or that a tracking email didn't make it back to you. It's up to you to maintain communications with your drop-shippers. Obtaining confirmation of shipments is vital, as is getting your orders processed and sent to the drop-shipper on time. The easiest way to solve shipment tracking is to have the drop-shipper use your shipping account (UPS, FedEx, AirBorne, U.S. Postal Service). This will enable you to receive all shipping and tracking information directly, rather than depending on the drop-shipper to supply you with the information.

Questions you should ask before delving into the world of drop-shipping include:

- How do you manage your drop-ship programs in order to ensure consistently superior customer service?
- How do you receive a confirmation that your customer has received the order?
- Are you alerted of potential problems (e.g. out-of-stock, shipment lost, etc.)? If so, how?
- Does the drop-shipper(s) provide tracking numbers for the products shipped? If so, how will you track each shipment?
- How are returns handled?

Some drop-shippers are slow to adopt new technology, so even if you are using cutting edge technology your drop-shipper may still use faxes and telephones as its main mode of communication, making it difficult to automate your logistics and fulfillment processes. If you find this to be the case, establish internal monitoring processes, assign employees to monitor drop-shipments, and leverage technologies that will help to integrate the website with the drop-shipper's systems to aid you in keeping the customer informed.

Another tack to take is rather than managing the drop-shipping process on your own, consider hiring a third party order management company to handle the task. If you choose this route, look for a company that will allow you to interface with their systems (e.g. a web-based tool that provides you with up-to-date information on product availability, shipments and customer deliveries). But know that while contracting with an order management company will make your life easier, the costs may cut deeply into your profits. Those web operators that may want to investigate the possibility of third-party help should check out CommercialWare.com, Dotcomdist.com, Netship.com, and Vendornet.com.

The confidence a website instills in its customers will be worth the struggle that it went through to provide the information. And your web brand will be paid ten-fold in ongoing consumer loyalty. A lack of shipping confirmations results in worried customers interacting with customer service representatives to obtain assurance that the order is on its way (an avoidable expense).

The "Holy Grail" of customer confirmation is to keep the customer in the loop by providing a confirmation that outlines the order's details with everything from the recipient's name, to the cost and billing information. Include:

* A personalized greeting.
* The order number and summary with charges.
* Estimated shipping and arrival date.
* Bill-to and ship-to addresses.
* Customer service contact information.
* Special request information (gift wrap/gift messages, special delivery instructions).
* Links to order status and order history.
* Links to return information.

The shipping confirmation should include all of the above information plus tracking numbers (when relevant) and links to the website's package tracking service.

Again, a website's customers expect a lot of information — ideally shipping costs, delivery times, order status, product availability, real-time online customer service — and they want easy access to it on the Web. Providing this information on your website — versus the fulfillment provider or the shipper providing the information — is preferable for branding purposes. All of this may require a large investment in what may seem like an overly aggressive web-based supply chain management technology, but as your web-based business prospers it will become a necessity.

✪ FULFILLMENT MODELS

The "back-office" subsystems of e-commerce sites — those that provide the link between the customer experience and the actual physical delivery of goods to the customer — are a challenge for web-based businesses. These parts, which include inventory management, order capture and management, and reconciliation, often prove to be more difficult than the construction of the site itself.

Online stores generally fit into one of six order fulfillment models, each with distinct benefits and faults. Let's take a closer look at each of these models.

The Single Website Order Model

The product is ordered off the website and shipped from an in-house fulfillment center (a basement or a garage for a home-based business; a warehouse or distribution center, possibly automated, for larger businesses) to the customer in one shipment. This model has one great advantage over other models — easy returns — since it gives the web-based business an increased level of control over every moment of the brand experience. This is what the mail order catalog business has been doing for years. However, it requires know-how, labor, facilities and often, special equipment and software, all of which may be outside the area of expertise and budget of the average web-based business. Also there is usually only the one shipping location, which means that same-day delivery is impossible (except in a small geographical area) and next-day delivery can be expensive.

Kiosk Order Model

Some brick-and-mortars use this model (and in the future you may even find pure-play web-based businesses in this space). The customer orders products, using a kiosk with a computer and high-speed connection located at a store or mall. With this model, your customers don't need a home computer and generally can even pay in person for the products ordered. This model makes it possible to order products online and pay in cash. Customers also have the choice of picking up the order on their next trip to the kiosk location or having it shipped. The big advantage of this model is a possible increase in customer loyalty — there is human interaction (the store clerk who takes the payment and hands over the products). Also, with a kiosk model, an almost unlimited range of products can be made available at a neighborhood store, provided the customer can wait a couple of day (if they are not willing to pay for next day delivery). Internet kiosks may become as prevalent as ATMs, so you can order your favorite books, wine, clothing, and gourmet food at your corner coffee shop or dry cleaners and pick up the order a few days later.

The Drop-Ship Model

This model is popular with some small e-commerce businesses, primarily because it reduces initial capital expenditures. With this model, a customer clicks the "buy" button on a website, the website then forwards the order to it's drop-shipper(s) (a wholesaler or distributor that owns a variety of products). The drop-shipper then fills that order by shipping the product directly to the website's customers.

The web-based business owns the customer database, while the drop-shipper owns the products. The drop-shipper typically pays the web-based business a sales commission on each product sold.

Here's how the drop-ship model works: You take orders on your website and you are responsible for collecting payments for all charges. Then at the end of the day, you put together a set of receipts and packing slips to forward to your drop-shipper(s). This can be done either by fax, email, EDI, or via the drop-shipper(s) website. The drop-shipper then takes the order information, packs a box, puts your receipt in the box, and ships it out. If you've done your homework and negotiations correctly, the customer won't even know that the drop-shipper exists.

However, before you take your first order, you must have a rock solid contract in place with the your drop-shipper(s). Sharpen your negotiation skills because every little detail must be put in writing or you may find yourself continually arguing over little things such as tacked on stocking fees, extra shipping fees, etc. Spell out everything in your contract including your payment terms.

The drawbacks of this arrangement are that there is little or no control over how the products are packed and shipped. Furthermore, unless you are diligent in your contract negotiations, your products may be shipped with the drop-shipper's name on the packaging instead of your brand. This can cause confusion and difficulty in retaining a customer base loyal to your website.

There are two ways to drop-ship. One is actually to buy the inventory and the drop-shipper charges you to store it and to ship it for you. The other is to work out a cost per item (plus fees) and then whatever you sell over that cost will be your commission or profit.

Those e-commerce operators that adopt the drop-ship model essentially reduce their capital risk in exchange for higher, overall operating costs. Here's why. If you offer products from more than one drop-shipper, the manhours it takes to manage these vendor relationships, to get the proper orders timely delivered to the drop-shipper, to make sure the product and service is good, and to update a sales and tracking system, can really eat into your profits. Also you must be careful not to let the drop-ship model obscure your band name or jeopardize your customer relationships.

The best way to adopt a drop-ship model is to start small with one company that has a good reputation and that offers a lot of products. Negotiate the best deal you can, make sure your deal has some sort of "out" clause in it. Once you have increased sales volume, you can come back and get a better deal.

Another downside to the drop-ship model is that it requires the e-commerce business give up much of its control of the fulfillment process. Although the e-commerce operator may establish rigid guidelines for the drop-ship suppliers, the business is still putting its brand into the hands of strangers. This includes everything from quality of product, delivery to the customers, and communication with customers concerning tracking, shipping, delivery dates, returns, and so forth.

One more disadvantage of this model becomes apparent when you try to integrate the front-end customer experience with the drop-ship fulfillment process. A good example is the shopping cart software. It's difficult to use most of the traditional shopping carts to gather your orders. Many plug-in shipping calculators and carts aren't compatible with drop-shipping needs. They can't deal with orders that contain products with varying point-of-origin zip codes. There are work-arounds, however. For example, look for a shopping cart that can be modified and then find a shipper that can provide a plug-in that can calculate live shipping rates based on different zip code origins of different products within a single order. Finally, if you are using a web-hosting service, the server needs a specific module installed on it to deal with the new shipping calculation technology.

If adopting this model, be sure to:

* Research shopping cart applications very carefully — the product you purchase must be able to be modified to your specifications as a website that uses drop-shipping.
* Determine whether you are limited as to what shipping companies you can use to deliver products to your customers.
* Keep control over when an order will be shipped to the customer and in how many packages.
* Determine how to keep track of what your drop-shipper has in inventory. The best way is to only deal with vendors that can provide you with a direct feed of current inventory; this way you can keep the products on your website current, avoiding backorders and out-of-stock problems. If that isn't possible, be proactive —know what products are in stock on any given day.
* Realize that it's up to you, not the drop-shipper, to provide your customers with up-to-date tracking or shipping status information.
* If using multiple drop-shippers, realize that they may want to receive orders in multiple formats (EDI, XML, email, fax, etc.). It's up to you to determine how this will be accomplished.
* Ensure that your brand is on the documents and packaging sent to the customer (e.g. packing slip, shipping label, box) — not the drop-shipper's information.

Keep in mind that during contract negotiations with a drop-shipper, your first priority is to protect your company, your brand, and your customers. However, if you are diligent and put in place procedures to maintain a firm upper hand, you have a reasonable chance of building and maintaining a satisfied and loyal customer base using the drop-ship model.

Same-Day Home Delivery Model

The products are ordered online, picked, and packed at a local distribution center, and delivered the same day to the customer. This approach offers the convenience of local shopping with an added benefit of being a timesaver. However, to provide same day delivery is always more expensive than other fulfillment models — both for the web-based business and its customers. Still, most customers that frequent websites that offer same day delivery feel it is a good tradeoff — personal time for money.

This model requires a very complex distribution and delivery network. It also is limited to specific geographic areas (how is a website located in Connecticut going to give same day delivery to a farmer in the outreaches of Minnesota or a customer in the mountains of New Mexico).

Fulfillment Service Provider Model (FSP)

With this model, the e-commerce business outsources its warehousing and distribution services to a third party. Companies that offer these type of services are called "third party logistics providers," "fulfillment service providers," or "fulfillment houses" (this book uses "fulfillment service provider" or "FSP"). This model gives the web-based business a good deal of control over all aspects of product quality, distribution and messaging. If you use this fulfillment model, you might want to consider contracting for more than one distribution center — each strategically located nationwide or worldwide, and each center carrying inventory levels relative to their regional market.

The FSP receives merchandise from not only your e-commerce business, but also numerous other clients and then stores the merchandise in its warehouse(s). It picks and ships the orders received by its clients. Since the FSP provides this service to numerous client businesses, it can spread the costs of the operation across a large base. However, web-based businesses may find that maintaining the optimum service levels require adjustments — give everyone at least four or five months to work out all of the bugs. In particular, many of these providers make it difficult for a web-based business to obtain real-time information about inventory status and order status. Don't sign a contract with a new FSP just before the Christmas buying rush.

Many large websites use this model. They feel (and rightly so) that, in the short-term, it's good business practice to concentrate on their core strengths and contract with experts for other areas, such as a FSP. The FSP will, for an up-front agreed price, provide whatever services are needed, e.g. the FSP will receive a website's products, warehouse them, and when customers orders are received, it will pick and fill the orders, and pack and ship each order by the method that the website and/or its customers choose. The provider works for the web-based business and therefore the labels and packing lists carry the website's name and logo.

The fulfillment service provider model is frequently identified as the future of e-commerce, but it's not there yet. While many web-based businesses give this model a try, many eventually find that the expense of fulfillment eventually outweighs its benefits.

In-Store Fulfillment Model

This is strictly a click-and-mortar model. When an online order is received by the website, that order is fulfilled by employees who pick stock from the traditional retail store's shelves. From there, the product delivery process is basically the same as any other model. This model incurs lower startup costs in the short term, which is why some brick-and-mortars use this model to "dip their toe" into the e-commerce arena. In the end though, this model can be very expensive due to the overhead and the complexity in tracking and pricing the same inventory for store and web sales. The tricky part is identifying and separating costs to each entity.

✪ IN-HOUSE SOLUTIONS

Controlling the entire customer buying experience, from beginning to end, gives an e-commerce business a competitive advantage. However, this also means that you must learn how to make a profit on each shipped package. Yet, the sad fact is that most e-commerce operators cannot quote their business's cost of fulfillment.

Demand for order fulfillment solutions is reshaping the existing e-commerce landscape. While most web-based businesses initially survived by offering a limited number of products, to compete most have been forced to expand their product range.

Any e-commerce business with more than 1000 orders per day should bring its e-commerce fulfillment in-house for three reasons:

1. It gives more control of the operations.
2. It leverages economies between existing and online channels.
3. It allows differentiation of customer service.

The Small- and Mid-sized Website

The author understands that the small- to mid-sized e-commerce business can't afford the infrastructure that the enterprise web-based business puts in place, and for these smaller entities it's not absolutely essential. That's because they don't have a large inventory to deal with daily. Remember your logistics plan? Use it to give you a good idea of what you need to do.

Order processing can be tamed if the volume is within a manageable range. As orders come in, send email acknowledgements to the customers. Then, process the orders through the payment authorization process, generate invoices, and log all information into the accounting software and inventory software. With such a system a website can know how much inventory has been sold, to whom, and at what price.

Once the initial processes are completed, it is time to pick the orders (find the physical products in your inventory), package them along with packing lists, invoices, return information and return labels, and any promotional materials.

Now the pick-ups are ready to be delivered to the shipper. Once the packages are in the hands of the shipper, send follow-up emails to the customers with the total charges, shipping information including tracking information (if available), and estimated delivery date.

Once the package is delivered to the customer (and it is up to the web-based business to keep on top of this), send a quick email to the customer to verify that the delivery was received and that the product(s) met the customer's expectations.

Software, etc.

Inventory issues can lay waste to a website that is "starting to take off." To alleviate this problem, many e-commerce software packages provide a full-featured inventory control system, including out-of-stock and low-stock notification by email. If the e-commerce software your website uses doesn't offer this feature, see if an upgrade is available. Or your web-hosting service may offer an online inventory system that you can use to synchronize your offline inventory system, which may be in the form of a flat file database, Corel Paradox, Microsoft Access, Microsoft Works, etc. If using the latter system, test your web-hosting service's order-activity file formats to see which has the least compatibility issues.

Another suggestion is to use off-the-shelf software like Quicken QuickPOS, Intuit's QuickBooks, PeachTree Accounting, or Microsoft Excel since all have good inventory management packages which can help small websites manage their inventory and track their orders. Finally, a small but active website might want to take a look at Hallogram's IntelliTrack Data Management Software (www.hallogram.com).

Whatever inventory system you use, carefully monitor the sales activity of each product in your inventory and set realistic reorder levels (so the website doesn't run out of its best sellers). If you do run out of a product make sure your system can place a "temporarily out of stock" notice adjacent to that product with approximate replenishment date. Also look for a system that allows you to insert additional information (cross-selling) about another product that may "fill the bill."

If the website's volume of orders is such that additional automation is advisable, help might be available through products such as Bonafide Management Systems' eResponse (www.bonafide.com) order fulfillment software. This product can be useful in all type of e-commerce situations, from the small startup to the large, high-volume e-commerce site.

Another system worth a look, especially for small- to mid-sized website, is Order Desk Pro (www.odpro.com). This order entry, order tracking, sales performance analysis, and customer database management tool is specifically designed for companies that receive orders via a website.

A busy, mid-sized website might want to investigate Verian Technologies's(www.procureit.com) ProcureIT order processing and inventory software that works with an ODBC database.

Shipping Procedures

We will first discuss the customer's view. Will you offer shipping options to your customers? Will the website offers multiple shipping methods? If so, give your customers the tools to pick their shipper of choice and the method (i.e., overnight, 2nd day, standard) and calculate the cost of shipping. A good practice for small websites is to offer links to each shipper's site on the web page where the shipping information is provided. In this way the customers can calculate cost and track their package (assuming it is sent with a tracking number).

Your shipping page should include the website's shipping policies set out in a clear, concise manner, sample weight/price shipping charts, and perhaps a note about the amount of time that the average package will spend in transit. Educate your customers. Make them aware of all shipping options. Many will opt for second-day service when they realize the difference in cost.

Now we will look at getting the product to the customer. Check with the customer service representatives at the larger courier services, including the U.S. Postal Service (if your business is located in the U.S.). Ask about rates and services for new business accounts. Compare times for last drop off and delivery times. Ask if they offer guaranteed delivery, refunds, insurance, and free supplies.

The U.S. Postal Service provides Express Mail Corporate Accounts, which are a big help since they let patrons avoid weighing packages, affixing stamps, and battling with postage meters. To set up an account, complete the application on the Postal Service's website (http://new.usps.com) and either mail it in or take it to a local post office with the initial deposit.

Note that because of heightened security restrictions by the Federal Aviation Administration, any domestic mail other than Express Mail weighing 16 ounces or more and bearing stamps must be given in person to a retail clerk at a post office or given directly to a letter carrier.

The U.S. Postal Service, either through its website or at the local post office, offers many of the supplies needed by a fulfillment center. For example, Express, Priority, and Global Envelopes, rolls of Priority Mail tape, sturdy boxes for shipping fragile items, padded envelopes for small items, pre-printed Express forms and self adhesive Priority Mail labels with the web-based business' return address. It is also noted that the U.S. Postal Service offers what it calls an "e-merchandise return" service — check it out.

Courier services including Federal Express, Airborne, and UPS, make it easy to set up an account online. Most courier services will schedule daily package pick-ups although some may charge a small weekly fee for pick-up service, others will do it for nothing (especially if you have an established business account with them). Still, if your shipping facilities are located near any of the major couriers' drop-off points, or near one of their drop boxes, it might be easier to just drop off the packages.

FedEx offers several software packages to ease your e-commerce business's shipping chores. **Airborne** offers a number of software packages to help you to better manage shipping products through its service. **UPS** offers its Online Office software to help web-based businesses process shipments, print address labels and pickup records, and track packages. **DHL** has Easy Ship software that it provides to businesses that meet specific criteria.

Negotiate with the shippers — it can't do any harm and most of the time it is possible to obtain volume discounts, especially if the rates offered by their competitors are brought into play. Then pass on any savings obtained to your customers. As orders increase, re-open the negotiations. It is possible to optimize your website's shipping solutions by exploiting each shipper's strengths while getting the lowest rates possible.

Insure all packages that contain items with a retail value in excess of $100 should be insured. The standard insurance provided by couriers usually cover the first $100. Simply fill out the corresponding space on the air bill — there will be a small fee charged for each additional $100 beyond the first $100.

Stay on top of the delivery time of all shipments. If a certain courier offers a guaranteed time (or day) of arrival, and that guarantee is not met, ask for a refund. In all probably the website will have a disgruntled customer when a package arrives a day later than expected, and your business will be the whipping boy, not the shipper. Passing the refund on to the customer can assuage that anger, while building your reputation as an e-commerce site that offers outstanding customer service.

Large and Enterprise Websites

As stated previously, for a brick-and-mortar making its move to the Web one of the

EXPLANATION OF BACK-END INTEGRATION STANDARDS

Application Programming Interface (API): Software that an application uses to request and to carry out lower-level services performed by a computer's operation system. In short, an API is a hook into software. As such, an API is a set of standard software interrupts, calls and data formats that applications use to initiate contact with network services, mainframe communication programs, etc. Applications use APIs to call services that transport data across a network.

Electronic Data Interchange (EDI): A series of standards that provide computer-to-computer exchange of business documents between different companies' computers over the Internet. EDI allows for the transmission of purchase orders, shipping documents, invoices, invoice payments, etc. between a web-based business and its trading partners.

eXtensible Markup Language (XML): A standard system for organizing and tagging elements of a document. XML has the ability to structure exchanges of data between computers attached to the Web, thus allowing one web server to talk to another web server. This means manufacturers and merchants can quickly swap data, such as pricing, stock-keeping numbers, transaction terms, and product descriptions.

first steps is to integrate its back-end systems. Until now, website integration issues were limited to the applications residing within the web-based business's own domain. However, logistics technology requires that integration now extend outward toward the four corners of the world and application program interfaces (APIs) are the passports. Most e-commerce systems have APIs but they vary widely in capability — legacy systems have APIs with varying capabilities and there are even custom-built legacy systems without APIs (but it still might be possible to modify these system to directly use the website system's APIs).

If taking the API route is too pricey, risky, or inadequate, the next course to take is Electronic Data Interchange (EDI) or Extensible Markup Language (XML). Both provide integration between systems. EDI is a well-established standard — the EDI messages are normally transmitted between businesses by Value-Added Networks (VANs). XML is an emerging standard for describing information and can be transmitted directly between businesses over the Internet, but translation will be required between the legacy system and the e-commerce system.

The Options

After evaluating its present back-end systems and their current needs, an e-commerce site usually has the choice of three options for its e-commerce platform:

A Custom Solution: This is a good choice, but only if you have the capital and the time to design a comprehensive solution to give your website the integration it needs.

A Packaged Solution: This can, at times, be less pricey than a custom solution. A packaged or product solution is not the typical "out of the box" product (this solution can't be bought in a retail store). While less costly and time-consuming than a custom solution, a packaged solution still requires specialized integration, i.e. there is an additional expense for "professional services" and it will take time to get everything working smoothly. If considering this type of solution, do your research to find the right packaged solution that can integrate with your business's existing systems since not all APIs are equal. Also determine how the packaged solution's APIs will operate with remote systems and with the website's specific fulfillment needs and existing systems. Finally, don't forget to check out how the solution handles or affects your systems' security features.

Many high-end packaged solutions offer API, EDI, and XML capabilities. But it pays to be careful, a product that states it offers EDI capabilities could only mean that it simply sends an EDI850 purchase order message in response to an order. This falls short of the kind of meaningful interchange needed to support a website's fulfillment needs.

Finally, if the existing e-commerce system is currently running on a packaged product, it may be that the necessary usable hooks and tools to integrate the product will cost many times more than the original cost.

Application Service Provider (ASP): It seems every day a new ASP offering a specialized service opens its doors for business. There may be an ASP out there calling your name. In fact, ASPs may be a good alternative solution for some websites since ASPs maintain software and infrastructure for many clients, the cost can be spread across the landscape, resulting in lower costs for its clients.

The ASP model works because the costs — software and integration — are often built into a long-term contract and amortized. A good ASP can provide an e-commerce solution that seems like it is customized but without incurring the huge development, maintenance, and infrastructure costs.

You can find ASPs that offer third party software and/or proprietary software. If the ASP offers third party software then there might be a few complications: The ASP may have built-in cost factors that could limit the value proposition to a website. The ASP is forced to rely on the third party vendor to fix bugs or to provide answers to technical problems.

An ASP that offers its own proprietary software usually provides a lower ongoing cost factor and greater flexibility. These ASPs also seem to be more flexible in their pricing policies. A website also might find this ASP model to be more responsive to adding product features or fixing bugs.

The shared infrastructure of the ASP model also provides architectural advantages, which can reduce the cost and time required for setup and integration. Just be sure that your IT people go over everything with a fine tooth comb to verify that the ASP can meet the website's specific needs.

Once everyone is satisfied with the decision — be it custom, packaged or ASP — get those legal eagles involved and set it down in writing how integration is to be addressed, what the costs will be, and how long it will take. Get warranties and guarantees with penalties for overruns.

Note: Another consideration for brick-and-mortars: From an economic standpoint, many traditional brick-and-mortar businesses should keep their web-based business separate from their traditional business to enable them to track actual costs — especially if the web-based business experiences early losses.

Upgrading Fulfillment Operations for E-Commerce Activity

Although customer-centricity is the key for the rush to upgrade fulfillment capabilities, some websites are wondering whether they can simultaneously update their fulfillment process and avoid compromising the customer experience. The astute e-commerce operator knows that to maximize efficiencies and benefits it cannot afford to make fulfillment an either/or decision. What you could do when selecting the methods and tools to optimize the fulfillment process is to not integrate the entire customer experience *immediately*. However, each component should be fully knowledgeable of the other, otherwise response to the customer's needs won't be personalized, and delivery won't made to the customer's satisfaction.

Even large e-commerce businesses find it difficult to master direct-to-consumer fulfillment. It takes a lot of guts to set up your own fulfillment center, especially since most e-commerce sites make the majority of their gross income during the November and December shopping season. If you haven't done everything right, your fulfillment operations will fail under the holiday crunch. Nevertheless, a huge plurality of the large and enterprise websites (mainly click-and-mortars) are adamant about bringing fulfillment processes in-house.

The deciding component of a website's success when building an in-house fulfillment system is its willingness to view the entire fulfillment process as a realm without borders. When optimizing the fulfillment processes the web operator must examine everything from the customer's initial product order through to delivery to the customer's door. This means a website cannot restrict its view to internal operations alone.

Another problem with building an in-house system is the lack of a "single" solution to integrate all of the back-end order processing operations — from finding products from alternative suppliers, to expediting orders, keeping customers informed, through to the financial end. Even with a good logistics plan, all of this involves a mess of prickly specifications:

* Taking care of credit-card authorizations.
* Routing one order to multiple suppliers.
* Routing status updates from suppliers to customers.
* Handling order cancellations and product returns.
* Keeping up with the availability of inventory and the forecasting thereof.
* Shipping specifics.

And that's just the beginning; the list goes on and on.

The main dilemma is that all of these applications have no *de facto* communication standards. One solution might be to bring an extensive library of pre-built connectors into play; the connectors could reside at the website's hub, speaking the language of and translating between the various applications that must communicate with each other. Of course, this means that such a comprehensive integration project may be a long, resource intensive process. Most websites will find that the complication of building out customized solutions is their only alternative.

Some website operators also may realize that developing the needed expertise internally is not cost-effective. That group may decide that the faster, more efficient option is to contract with one of the new breed of consultants and software vendors who can help a web-based business forge a trail through this technological briar patch.

Before taking the upgrade route, you must first assess your business's current fulfillment needs and capabilities and compare that with your long-term goals. Then, determine what is missing in terms of operational and functional processes, strategic direction, technology and organization.

Once everyone has grasped what needs to be done and what can be done, the task then is to set out the tactical steps to fill in the missing pieces. Then begin the move toward complete implementation.

It's easy to agree upon a strategy and to identify what is missing but the sticky wicket is defining the tactical operational requirements. If a web-based business can set out a feasible plan that cuts to the nuts and bolts of the process, then it has a fighting chance. E-commerce operators that are willing to do their homework will have a better chance of success.

Once work has begun, everyone on the team should be available for daily or thrice weekly conference calls to avoid duplicated effort and to assure that everyone is in adherence to the logistics plan and timetable(s). Everyone must stay focused on the unequivocal requirement for a system flexible enough to grow over time.

Use an Existing Direct-to-Consumer Operation

When a brick-and-mortar is planning a move to the Web, it assumes the easiest and most economical route is to keep order fulfillment in-house. But as its web-based business grows, it must come to grips with the fact that its distribution system is based on pallets or many-item orders, and 9 times out of 10 it finds that it is struggling to handle the one- or two-product, small-package shipments that a web-based business encounters.

One option for enterprise websites choosing to keep their fulfillment in-house is to look for either a direct-to-consumer competency buried somewhere in-house (such as a small catalog operation) or buy a logistics firm or a competitor with e-commerce fulfillment capabilities already in place. Either alternative can make it easier to implement technologies to handle basic e-procurement or elementary e-commerce tasks.

However, this course won't give a web-based business complete integration of internal and external systems across the entire value chain. Click-and-mortars that want to integrate their systems to support real-time two-way flow of information throughout the value chain (the entire supply chain: customers, internal applications, and financial) will need require additional custom solutions.

FULFILLMENT: IN-HOUSE OR OUTSOURCED

A PriceWaterhouseCoopers and eRetailing World study looked at how e-commerce businesses handle their logistics and fulfillment chores. Here is what they found:

E-Commerce Startup	Inventory Warehousing	Picking/ Packing	Shipping	Returns	Replenish- ment
In-house	47.2%	41.6%	36.1%	63.9%	52.8%
Outsourced	41.7%	44.4%	47.2%	22.2%	25.0%
Combination	8.4%	11.2%	13.9%	13.9%	16.7%
Other response	2.7%	2.8%	2.8%	0.0%	5.5%
Click-and-Mortar					
Company handles	71.8%	69.2%	66.7%	79.5%	76.9%
Outsourced	20.5%	20.5%	23.1%	12.8%	12.8%
Combination	5.1%	7.7%	7.7%	5.2%	7.7%
Other response	2.6%	2.6%	2.5%	2.5%	2.6%

As the figures show, although there are benefits of keeping fulfillment activities in-house, outsourcing is still a popular option for many e-commerce sites.

✪ OUTSOURCING

The goal of outsourcing fulfillment should not be one of short-term cost reduction, but rather speed to market. That is because in the long run, outsourcing usually costs more than an in-house system. The benefits of using a FSP is that it provides a reliable method of getting the product to the customer while keeping in-house control of important data such as inventory, reorder levels, and delivery confirmations. The other advantage of outsourcing fulfillment is that it allows a web-based business to focus its energy and resources on marketing, product development, sales, and building its customer base.

We've discussed outsourcing throughout this chapter; this section is where we really dig into the subject. While outsourcing can make fulfillment and distribution simpler and more efficient it does have limitations and drawbacks.

* Outsourcing is not cheap.
* Many FSPs will not service a website unless it has reached a decent shipping volume.
* An inventory that consists of a large variety of products can affect service and price. For example, it's more costly and time-consuming to prepare packages that contain different combinations of items than it is single-product packages.
* Some FSPs will only respond to inquiries regarding the status of an order while others will provide full product support.
* The level of services in warehousing and inventory management also may vary from company to company. In some cases, the FSP will order stock directly from the manufacturers while others will expect the website to ship the products to the FSP's main warehouse for packaging and delivery.
* The drop-shipment option may result in multiple shipping charges for some orders.
* Websites using drop-shippers will have to deal with customer confusion due to packages arriving at different times from different shipping locations sometimes without the website's brand anywhere in site on the packages.
* The drop-shipment option requires complex tracking and returns procedures compared to FSP shipments.

When Should You Outsource?

Outsourcing to a FSP offers quick access to proven state of the art technologies and thus can be a good starting point for a new web-based business. Lower start-up costs preserve scarce start-up capital. These initial savings however mean a higher per product cost during the operation of the website. Then there is the concern that any outsourcing model presents: How to keep control of the management of quality, accuracy, accountability, and customer service priorities, as well as integrating it into your back-end systems.

*Outsourcing to a FSP **only** makes good financial sense for a startup website if, in its day-to-day operations, it has **more money than time**. When every penny counts, keep fulfillment in-house. If the web-based business' competency lies elsewhere and it can earn more utilizing its staff in their areas of competency, look outside. If an established website currently handles fulfillment in-house, it should look at outsourcing **only** when it has **reached the limits** of its current facilities and infrastructure and the incremental cost of expansion would be a prohibitive expense or it can't have the disruptions that a rebuild of its infrastructure would bear on business.*

A web-based business might consider contracting with a FSP for the same reason it chose to outsource its website to a web-hosting services — because an FSP can afford the very best software and hardware solutions, and because its costs are spread over many customers. Thus, even though there is an additional cost involved with out-sourcing, it can give a web-based business the additional benefit of having a cutting edge fulfillment process at its fingertips.

What an FSP Offers

The average FSP offers everything a web-based business might need for its fulfillment processes. These services will take the ordered products from the warehouse shelves, pack them, and hand them to shippers. Then they will follow-up by sending an auto-mated email response with the website's branding to the customer to let them know the package is in transit. Many also will handle your credit-card processing, supply current inventory levels to the website, reorder products, offer call-center services, send notices of shipping, and handle returns. There are literally thousands of these companies to choose from, but the best way to find one that suits your specific e-business' needs is by referral.

Some FSPs offer everything: web design and hosting, order capturing, shopping cart tools, credit card processing, merchant accounts, picking and packing, shipping, invoice generation, inventory management, return/exchange logistics, and customer service. Plus some will provide, for an additional fee, extra services such as wrapping gifts and packing catalogues or other promotional items with each shipped package.

It sounds good, but remember some of these areas may not be among the FSP's core competencies and may not be the best solution for your web-based business. Care-fully evaluate your needs, the FSP and what it offers, and then pick and choose.

Most FSPs will let an e-commerce business mix and match the fulfillment services it needs. For example, if a website already has a shopping cart and credit-card function, it can limit its selection to inventory warehousing, pick, pack and ship service, returns processing and customer service support. An e-commerce operator may want to use a combination of fulfillment options. For some websites, it may be more cost effective to handle the processing of small items in-house and only outsource the bulkier items. Other websites may have some of their products handled by drop-ship methods and others by the FSP.

Should an e-commerce business choose a small or larger FSP? Each offers advantages and disadvantages. A small local FSP may be best, especially if it makes it easier for quick replenishment of inventory or because it offers more personalized service. A larger national FSP, on the other hand, can give a nervous startup confidence that its customers will be adequately served throughout the United States.

Another, good reason for choosing the smaller FSP is that an e-commerce business could have its fulfillment services in numerous regional locations allowing it to emulate a just-in-time inventory model (discussed later in this chapter). Having more than one FSP also can alleviate concerns about relying on a single fulfillment resource (i.e., redundancy). There are many options. Discuss them all with each potential FSPs until you find a provider and a solution that fits your business's specific needs.

Where and how can you find fulfillment service providers? Begin with Yahoo!'s directory. To reach your goal click on the following: *Business and Economy > Business to Business > Marketing and Advertising > Fulfillment Services*. The list should provide you with at least four or five companies that suit your needs.

Drop-shipping

In a perfect world the drop-ship model would be the way all e-commerce sites would want to go. If everything worked flawlessly, the website would concentrate on sales and therefore sell more products, which in turn benefits both the website and the drop-shipper. The website does not have inventory cost and the drop-shipper does not have the marketing costs; this enables both to make more money by reducing the retail price. But it's not a perfect world.

Although many small and home-based websites may go with the 100% drop-ship model when first opening shop, most gradually move into a hybrid solution — they still have relationships with drop-ship manufacturers and distributors but they also have begun to amass an inventory and deal with in-house fulfillment processes.

The drop-ship model has been discussed throughout this chapter, but as explained earlier, a website that uses the drop-ship model takes orders and payment for product and forwards its orders to the drop-shipper who will "drop-ship" products directly to the website's customers. When using a 100% drop-ship model, the website realizes some advantage from delegating the physical labor of the entire fulfillment process — from the order and stocking of product through the picking and shipping of the products to the customers — although as the reader should understand, the website does not totally avoid fulfillment expenses. That's because the e-commerce operator is compelled to be particularly diligent to guarantee that the fulfillment process is completed to the customer's satisfaction. If you fail in this task, you will lose control over your fulfillment process — you won't know when orders are shipped, or how they are shipped (including the number of packages, shipper, and method).

Note: Another challenge that the drop-ship model presents to e-commerce operators is managing, synchronizing, and consolidating customers orders so that all packages comprising any one order arrives at the customer's door on the same day by the same shipper.

Regional Distribution Centers

Suppose a FSP ran a network of distribution centers, and each center could ship product for dozens of large and small e-commerce sites. A web-based business could contract with one of the FSPs with multiple distribution centers, giving it a presence in many regional areas.

Using the regional distribution center model allows a web-based business to use just-in-time (JIT) inventory management, thus saving on inventory holding costs. With the JIT inventory model, vendors agree to deliver products right at the time needed — not so early that you wind up with unnecessary warehousing expenses, but not so late that you miss out on customer sales. Since needed products arrive at the warehouse just in time for shipment, you can keep your inventory at a level that is just enough to cover ongoing orders but not tie up capital.

Just-in-time inventory management and delivery works best with companies that have projects that are carefully scheduled and effectively managed so that work is performed according to a strict schedule. For example, construction companies may effectively use just-in-time delivery so that goods and materials arrive at the construction site just in time. Money is saved largely by not having to provide warehouse facilities for these goods and materials.

For websites that have substantial warehousing costs and large well-scheduled projects, just-in-time delivery may make sense. Those companies should ensure that there is no premium in the price for such delivery — or, if there is, that the premium is more than offset by the savings in warehousing costs.

When a business has only one distribution center its products are usually received in large quantities and stored in pallet racks until needed. As items are picked, more products have to be dropped from the pallet racks to replenish the picking area. This receive-store-replenish-pick cycle is a repeated expense, and, if replenishment is too slow, out-of-stock notices may occur even when product is in the building.

If a website uses the JIT model and outsources some or all of its fulfillment needs, it should look for a FSP that has supply chain management software to accommodate just-in-time shipping; thus allowing products and orders to flow smoothly from manufacturers and distributors to regional distribution centers and finally to your customers. Adoption of a JIT model can result in increasing the frequency with which inbound shipments of product are scheduled, but decreasing the lead times and size of these shipments. Web-based businesses that adopt the JIT model often reduce the number of suppliers and transport companies with which they must deal and select suppliers that are close to their regional distribution centers, enabling delivery of shipments with short lead times. In addition, each regional distribution center has only a portion of the website's total stock, so it is a simple task to shelve all received product directly into the pick area.

Another reason for choosing this distribution model is that many web-based businesses want to give their customers reasonably priced next day delivery or 2nd day delivery. A website could place its inventory in numerous regional centers, thereby increasing its inventory holding costs, decreasing its overall shipping costs, and increasing its customer satisfaction ratio.

Here's how it works. A business has one distribution center in (let's say) Nashville, Tennessee and a customer in Oregon who places an order for 2nd day delivery. Normally that order could be shipped for delivery within the two-day time period. But, what if there is a problem (which happens frequently). Now the business has to ship "next day" at additional cost (which it has to "eat") to meet the 2nd day delivery time limitation.

Now take that same scenario, but the business has its inventory in four regional distribution centers. The Oregon customer's order is sent to a regional distribution center in Utah; and, even if there is a mix-up which causes a delay in shipping the product, it can avoid the "extra cost" next-day shipping method due to the proximity of the distribution center to the customer. In other words, with strategically located distribution centers throughout the country a website can promise 2nd day delivery with more certainty due to the proximity of the product to the customer.

It isn't always necessary to increase inventory. With the right technology running behind the scenes, a web-based business can maximize its normal inventory levels.

Take the same scenario: the Oregon customer's order is sent to the Utah distribution center, but that center is out of stock and cannot fill the order. The fulfillment system can transmit the order seamlessly to another distribution center and still meet the 2nd day delivery deadline by using the more expensive "next day" shipping option. The web-based business did not carry extra inventory, but because of its data network it could still adopt the just-in-time model. Using this system, premium shipping charges would still be incurred to meet the 2nd day delivery deadline, but not in every instance, every day.

Your Outsourcing Plan

Nothing succeeds without a plan — this time it's an outsourcing plan, written as an addendum to the logistics plan. First, determine the services the website needs and expects from a fulfillment provider and what they will need to know about the website — for example, its expected order volume, its business' terms, and whether it'll pay handling on a per-piece-picked basis or on a package-shipped basis.

FSPs differ widely on the variety of options they provide and the type of client they serve. Finding the right fit for a web-based business requires in-depth research, which includes asking for referrals. Once the research is completed and the information compiled, it will be possible to price out the most cost effective and efficient method of getting the products to the customer while maintaining an adequate profit margin. But it's not an easy decision.

The first step is to look at your carefully drawn up outsourcing plan. If the solution is still unclear, consider the following while keeping in mind that fulfillment is the last form of customer contact — if a website fails to embrace order fulfillment with the same vigor as it embraced online selling, it will experience not only distribution headaches, but also, customer defection.

Look at the current volume of orders and project what the volume is expected to be in six months, one year, three years and five years. With figures in hand, take a long hard look at the website's staff — can they handle the workload. If the website takes off and workload exceeds the staff's capacity, its customers won't be happy with the poor service that they may receive from a harassed and overworked staff. The web-based business may have jeopardized its hard-earned success with the wrong fulfillment decision.

Also, consider the web-based business' core competency, fulfillment may not be one of them. Stay on top of fulfillment issues, and as the business grows, re-evaluate. Before making your decision, get a pencil and paper out and compare the costs of adding more staff and facilities (including automation) with the cost of delegating the fulfillment process to an outside company.

If the website is still in the first growth period and has expectations of a huge growth spurt within the next six months — outsource. However, if the grow is projected to increase steadily over time and the in-house staff can currently adequately manage the fulfillment process, then there is time to build a proper in-house fulfillment infrastructure that can scale as the order volume increase.

A start-up website that expects to handle a substantial volume of orders from the get-go and does not have a warehouse and staff in place might find that in the beginning it could be more cost-effective to outsource its fulfillment processes to a FSP. In that case, the web-based business should ensure that its back-end is built so that it is scalable enough to accommodate a move to an in-house fulfillment processes when the need arises

If a website's order volume is low and inventory requirements are not onerous, then it is probably better to handle fulfillment in-house until the volume justifies a more formal order fulfillment procedure.

Your Product. Look at the products offered for sale via your website. If the answer is "yes" to either of the following two questions then the decision to outsource is a no-brainer.

1. Do your products require special handling, i.e., fragile, sensitive to temperature changes, or require special licensing (alcohol, perfumes, pharmaceuticals, etc.)?
2. Will the majority of the product orders include multiples of different products? This makes the picking and packing more labor intensive.

With a yes answer, a FSP should be found that will accommodate the specific product needs and still allow the e-business to earn a profit.

However, if it isn't possible to find an FSP to provide the special services required, an e-business has two choices: to outsource all fulfillment except the products requiring special handling, which will be handled in-house, or to keep all fulfillment processes in-house.

Customer Service. FSPs are well aware of customer service issues and most offer services that integrate well with the typical website infrastructure. Some of the customer service issues to be considered when instituting a fulfillment process include:

✴ Managing inventory (so the customer is confident the product is available and the website can cross-sell, if necessary).

* Offering multiple shipping options.
* Tracking orders (so the customer can stay informed).
* Handling returns and disputes.

Control. If a website decides to outsource with an FSP, then the FSP must become an integral part of the web-based business's processes. As such, it is important that everyone coordinate their business processes and infrastructure to maximize the benefits of each business' strengths.

Integration. A web-based business will need to send its customers' orders via email or FTP directly to the FSP's warehouse facility. The optimal FSP is one that offers software to enable a website to integrate its systems with the FSP's back-end systems so that there is real-time inventory information. The FSP should offer continuous technical support to assure that the operation runs smoothly from both ends. Bear in mind that the FSP selected will become a partner in the value chain, and as such, an essential arm of your e-commerce operations.

Issues Specific to a Small Website

It is possible to find FSPs that cater to the small business, allowing even the busy home-based website to take advantage of the latest cost- and time-saving techniques, such as supply chain software, and orders sent directly to the distribution center floor for picking.

Hence, a small web-based business that chooses to outsource fulfillment, should select a FSP that specializes in servicing small businesses, especially if it receives fewer than 10 orders per day. Do the due diligence — if the FSP falls down on the job during a peak selling season, a small website might not recover and could lose its entire business.

For the majority of small website's the cost per order is significantly increased when an FSP is brought into the picture. Outsourcing makes good financial sense ONLY when you have more money than time. If you need to make every cent count, handle your fulfillment in-house.

A series of coordinated steps is required to bridge the gap between (1) a customer clicking the buy button, thus sending the order to the shopping cart software and into the "pipeline" and (2) the customer receiving the products. Ad-hoc solutions to handle the communication with a FSP and/or drop-shipper(s) can be put into place that may be sufficient for a website with low sales volume and limited inventory. However, if sales increase substantially, then even a small website must deal with the lack of

EVALUATING THE OUTSOURCER

Keep in mind that an when you partner with an outsourcer (FSP and/or drop-shipper), it must become an integral part of your website's operations. Therefore, find an outsourcer that not only offers the services the website needs, but also is a comfortable fit in other ways — location(s), technology, mission statement, financial wherewithal, etc. To determine whether a prospective outsourcer meets a web-based business' needs, ask it how it approaches key fulfillment issues, such as:

- What types of products does the outsourcer have in stock for other clients?
- What is the minimum volume the outsourcer will handle? Is that minimum per month or averaged over a set period?
- How does it handle the packing? What type of packaging materials will be used? How will the box be labeled? In other words, is the website's brand used or also the outsourcer's brand? It's important, that the website's brand isn't diluted.
- Determine that the outsourcer does not require an exclusive contract to fill all of the website's product orders. The web-based business may want the flexibility of processing some of its orders in-house and also it may want to hedge its bets with a secondary FSP or additional drop-shippers.
- Is there a setup fee? If so, how much and is it a one-time fee?
- Does the outsourcer offer special services such as gift-wrapping or sub-assembly? If so, what are the fees? Will the outsourcer include the website's catalogs or other branding materials or special offers in the shipments? If so, is there an extra fees and what is the fee?
- How quickly will orders to be filled? Is there a guarantee?
- What volume of orders can the outsourcer reasonably handle and how scalable is its capacity?

standards for data representation and transmission requirements. This means dealing with the API issues or taking the EDI and XML approach (discussed earlier in this chapter) for communication between systems so as to be able to interact with each outsourcer (FSP and/or drop-shipper). Eventually, the "string and sticky tape" method will fail and proper integration will become essential.

How Much Does It Cost?

The cost is dependent upon which services the web-based business requires and the volume of orders it forwards to the outsourcer since many of the fees or commissions are set on a sliding scale based on volume. On the average, you will pay an FSP a monthly fee for warehousing the products of around three cents per small item, or $15 for each pallet of products. An inventory count is usually performed monthly and the fees are assessed on the products your business has in the FSP's warehouses at

- How will the website's infrastructure be integrated with the outsourcer's systems? Will special hardware and/or software be needed? What support does the outsourcer offer for the integration process?
- What credit cards and alternative payment methods does the outsourcer accept?
- How are sales taxes handled?
- How does the web-based business get paid and how quickly? (The drop-ship model.)
- How will the website and its customers track orders?
- How will the web-based business monitor and replenish inventory?
- What reports does the outsourcer provide?
- How will returns and disputes be handled?
- Does the outsourcer offer customer service support?
- Does the outsourcer handle international orders? If so, how? Are there extra costs involved?
- What shipping arrangement does the outsourcer have, and is there flexibility in that arrangement to allow for the use of other shippers if they have better rates?
- What kind of account servicing does the outsourcer offer?

Be sure to ask for references and a customer list of at least 10 customers. Call all the references and customers. Ask the customers (if you can get away with it) all the questions set out above, but at least ask what kind of products the outsourcer handles for them and how well the outsourcer meet its obligations to them. Ask them if they have any complaints or reservations about the outsourcer. Ask them if they plan to renew the contract when it comes due.

Another reason for demanding a customer list is so you can look it over to determine that there is any conflict between your business' product line and others that the outsourcer handles. The customer list will also alert you to any large clients that may compromise the outsourcer's resources, especially during a busy holiday shopping season.

that time. The FSP may also charge a handling fee for each package and for incidentals, such as mailing labels, boxes, packing material, and then, of course, shipping.

There are no outright, up-front expenses involved with the drop-ship model. The drop-shipper pays the website a commission, which can range anywhere from 10% to 30% or more of the product(s) sales price.

The Contract

Once you have decided upon an outsourcer, the next step is to perform due diligence concerning such issues as financial health, union related issues, stability of management, the customer list, references, and the physical plant (lease, ownership, condition of building and equipment, etc.).

If the due diligence turns up nothing untoward about the FSP, the next step is to get your attorney involved. Have him or her draft the contract, which should clearly outline both your and the FSP's expectations, services to be provided, fees to be paid, and guarantees. Give your attorney the logistics plan and the outsourcing plan to use as a guide. Keep in mind that both the website and the outsourcer want to maximize revenues, so it is important that everyone agrees on a compensation rate that is fair to all.

The contract that is signed should be for a limited term since outsourcing is a choice that you will reassess throughout the business relationship. However, the contract should be a concrete document that defines the services and responsibilities of both the website and the outsourcer.

After everything has been signed, the next task is integrating your website's systems with the FSP's back-end so that fulfillment processes are executed so seamlessly that your customers won't know the difference.

The Results

Now it's time for the outsourcer to go to work. If the FSP is handling credit-card settlement for your website, then the charge to the customer's credit card is authorized and settled for the order amount. The products are then shipped to the shipping address using the selected carrier. If a product requested by the customer is not available, the customer's order is placed on hold (backorder), until inventory is replenished.

During all of this the customer must be kept informed as to the order status. In addition, it is desirable that the customer be able to go to your website to bring up the order history and to request a real-time order-status query. If it's supported, the site communicates with the outsourcer via secure HTTP and requests the status. The response message should include details of all shipments, cancellations, backorders, and other transactions that have taken place while processing the order.

Most outsourcers send order status files that describe all of the day's order activity nightly, via FTP or encrypted email. By law, a business can't charge (settle) a customer's credit card until the order has been shipped. So, the order information returned to the website by the outsourcer(s) must be verified for accuracy before the credit-card information is processed.

Without valid information, the website can't post a shipment invoice to the order management system. Settlement tasks run periodically to process shipment invoices that need collection. The settlement process communicates with the clearinghouse to charge the buyer's credit card (note that when the order was originally placed, funds were only allocated [authorized] but not disbursed). If the site and the outsourcer(s) don't agree on order details, customers may be charged the wrong amount.

Then the information is transferred from the order management system to the website's financial system for purchase order generation, accounts payable, and so on.

Once the products are in the customer's hands, the website and the outsourcer jointly need to handle any post-purchase activity such as returns or exchanges, and to perform transaction reconciliation.

Caveats

Despite the many advantages of using a FSP and/or drop-shipper, there are a few potential downsides. Putting someone else between a website and its customer can be a risky proposition, and there are significant data integration issues. When sorting through these issues, it pays to see what others have done — get out there and talk to everyone you know who uses or has used an FSP and/or drop-shipper.

No matter what outsourcer(s) become your e-commerce fulfillment partner(s), issues will arise. What happens when the busy and ever important Christmas season arrives and it comes time to decide between shipping an order for a smaller company (maybe like your website) or shipping an order for Amazon? Which one is the out-sourcer going to choose? It's a given, the big guy will get the most attention.

A smaller web-based business can protect itself to some degree by establishing a detailed service-level agreement to ensure performance (get the legal eagles involved).

The Internet is a great equalizer. With the right FSP and the right integration between the FSP and the web-based business, all websites can do business on the same playing field with the same type of equipment and the same type of personnel.

Want to know more? Check out Opsandfulfillment.com (subscribe to the magazine) and read the author's book *Logistics and Fulfillment for e-business* (CMP Books).

✪ CONCLUSION

Regardless whether the e-commerce business is a start-up, a brick-and-mortar, a home-based business, small, medium, or large, all of the solutions discussed in this chapter do not come without cost. There are capital expenditures and development costs associated with integration, and channel conflict in regards to any type of consolidation.

Order processing and fulfillment is a website's last form of customer contact. Investing in a logistics strategy is critical to a website's continued growth. Whatever the cost and ultimate solution, it is clear that new methods and processes must be utilized for a web-based business to expect long-term survival in the e-commerce arena.

Chapter 19

Final Thoughts

You can only predict things after they've happened.

—Eugene Ionesco

Despite a few stumbles, the future is bright for e-commerce. The 20th Century, shaped by the Industrial Revolution, became the age of the automobile and the television. The 21st Century, shaped by the Technological Revolution, is the age of globalization. The Internet massively impacts all aspects of business. In the 21st century, e-business is no longer an option for businesses; it is a necessity.

Today, e-commerce is an ever-expanding consumer industry. For an e-commerce site to succeed it must understand its customers' mindset. Although price is always an issue, it is rarely the primary motivator for buying a product online. Customers are looking for convenience, and/or products they can't find elsewhere.

Small websites that cater to niche markets have the best chance of prospering. That is, as long as you take care to ensure that your customers' shopping experiences aren't marked with too many potholes.

✪ FACTORS THAT CAN SLOW GROWTH

The Internet, the Web, and e-commerce have their peccadilloes. External elements, including the public at large, and governments, have not yet weighed in with a final opinion regarding this new behemoth. The public's concern about security and the personal information that websites amass must be addressed before the public takes action of its own. In the other arena, a crucial and as yet unresolved factor that could impede the growth of e-commerce is the degree of governmental involvement including regulation and taxes.

Customers' Security Concerns

The general population believes the Web is an unsafe place to conduct transactions — specifically credit card purchases. But this perception will change as the customer gains more experience on the Web using sites that incorporate digital certificates, the SET protocol, and SSL. The technology is here — many websites use secure servers for online transactions. These servers encrypt data that is sent over the Internet so that a third party cannot intercept the information. Thus, many consumers will purchase products from a website when convinced that the technology is in place to safeguard their transactions and that the website itself is built upon a promise of integrity and trustworthiness.

Trust and e-commerce are mutually dependent. Although technology, such as using digital certificates, helps in the battle to instill trust, branding is the key. Your brand is the gauge a customer uses to assess quality and reliability. Stand by it and make it a symbol of your contract with the customer to provide not only a product but also service and satisfaction. Would you trust something if American Express endorsed it? Do you trust Lands End? Customers need the reassurance of familiar brands (new, old, traditional, and web-driven), and assurance that the technology is in place for a secure credit card transaction to build the trust required to take the first steps toward an online purchase.

Privacy Issues

The ability to guarantee the customers' privacy is an important challenge. The extent to which a website uses personal information concerns every Internet user. How much does the surfing public want the web community to know about them?

The growing trend toward personalization on the Web will have a considerable impact on e-commerce. While personalization does not translate into direct sales, it brings the customer one step closer to a purchase. If handled correctly, your customer may appreciate a familiar relationship that could include the recommendation of additional products or of timely notices of sales. But if you abuse it, you've lost a customer.

Still, personalization tools are becoming more prevalent. That's because, the better you know your customers, the easier it is to know what the individual customer wants. How you obtain the information necessary for this personalization is the issue. Some websites ask for information outright, and then use rules-based filtering systems or adaptive prediction technology that can provide real-time learning of customer likes and dislikes based on mouse clicks.

An information exchange is fair game — I answer the questions and then the website gives me a $5 coupon, a 20% discount, a t-shirt, etc. That website has literally become a storehouse of my personal information. If my information is passed on to another party without my permission, I take issue. This is where the e-commerce community must police themselves or the government will step in to regulate how websites use and exchange a customer's personal information.

Bear in mind that websites that utilize personalization tools, if managed properly, are rewarded as early innovators. The websites that have amassed personalized profiles can get to know their customers over time. The longer websites wait to adopt personalization, the greater head start their competitors will have in using personalization tools to gain a foothold in the customer base, making it unnecessary for those customers to seek out other sites. Personalization tools lend themselves well to the early adopters who have established a brand on the Web and who have focused on creating a long-term relationship with their customers.

✪ GLOBALIZATION

The Web is the first wave of a pure global economy. That means opportunity — to take advantage of a rapidly expanding global trade and the ensuing revenue stream. The task is deciding how a specific business can fit into the new global economy.

Many new websites receive at least one international order on the first day they are up and running — amazing proof that e-commerce generates a global customer base. Imagine from just one location, a business can reach a market anywhere in the world, not just in a specific state or country.

One barrier in this rosy picture is the average website owner's lack of understanding of worldwide taxes and tariffs. Integrating enabling modules into a commerce software package could be a first step in untangling this confusing issue. Also, websites need to keep costs down and ensure fulfillment efficiencies in the areas of tariffs, freight, and sales taxes.

All international e-commerce customers have the same questions - "If I buy this product from this site, will I pay a duty on it when it crosses the international border? What are the shipping costs? Can I avoid paying sales taxes?" If you are serving an international customer base, devote a web page to international customers with a complete, easy to understand explanation. Be prepared!

So, dear readers, go out there and fight the good battle. The Internet and the Web have an investment community and others who realize their value, and governments who right now are silently waiting in the wings to see how this new phenomenon shakes out. However, time is running out. Everyone wants a piece of the pie. Make your plan and carve out your piece of the Web — it's the future.

Appendix

Computer Basics

What is a Computer? A simple question with no easy answer — a conventional engineering definition is that a computer is first and foremost a tool, not an object, i.e. a computer is defined not by what it is but rather by what it does. But let's take a different course.

In the early days, computers were behemoth machines that carried out elaborate mathematical calculations. Programs were written to perform a desired calculation, submitted to the room-sized computer, and then you waited while the machine crunched your numbers and printed the results onto several reams of wide green-striped paper.

The next evolution raised the computer out of the big mainframe, batch-oriented, number-cruncher category into what were called microcomputers because these new machines were smaller versions of the older, bigger mainframes.

That brings us to the modern computer — a device that can be as small as your hand or as large as a small suitcase. It comes with a CPU, some RAM, an address bus, data bus, and so forth. The central focus of today's computers is interactivity. Their very essence is to explore. The human uses his or her creativity and values to propose a variation on an existing scheme. The computer uses its computational resources to work out the ramifications of the variation and presents the resultant order of things to the human. Today computers' responses are so fast that humans are encouraged to try again and again, with different variations, to explore all the possibilities, to play with each situation.

Now, let's examine the components that make up the average computer used to run a website. Situated in a computer's chassis are, at a minimum, the following:

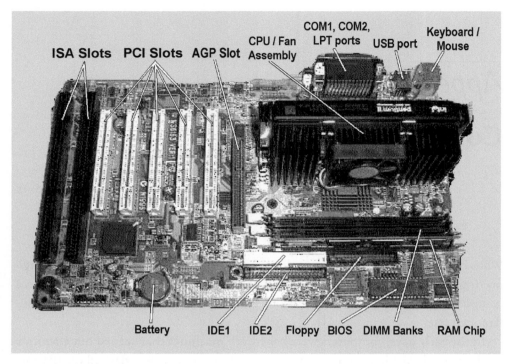

Figure 28. A typical motherboard.

The **motherboard** is a circuit board that everything plugs into and provides the actual physical connection between the different components that make up the computer.

A **system bus** connects the components on the motherboard and enables the CPU to communicate with all of the components. The motherboard consists of:

* **Memory** (RAM or Random Access Memory). These chips installed on single in-line memory modules (SIMMs), are a temporary high speed holding area for data and applications for fast access by the CPU.

* **CPU** (Central Processing Unit). This is a programmable device that processes digital information.

* **Chipset.** A collection of semiconductor chips that are the interface between the CPU and the rest of the computer, and instructs the computer how to organize and transmit data. The chipset will typically include such features as the system memory controller, the PCI controller, and the AGP controller.

* A special **graphics chip** to speed processing due to the complexity of graphics files.

Connected to the motherboard by cables are:

* The **hard drive(s)** offer large capacity and store most applications, including the operating system.
* **Floppy drive(s)** that are primarily used to load applications, and to retain and to transport small data files.
* **CD-ROM drive** (Compact Disc/Read Only Memory) can store larger applications and more data. The faster your CD-ROM the faster your software will load.

Plugged into the motherboard via the **AGP** (Accelerated Graphics Port), **ISA** (Industry Standard Architecture) and **PCI** (Peripheral Component Interconnect) **expansion slots** are various **adapter cards** that allow you to customize the computer, such as:

* Sound card (makes sound).
* Video card (shows graphics and pictures).
* Modem (to connect to the Internet) is an acronym for modulator/demodulator. It sends digital info through analog telephone lines.
* Network card (to connect to other computers).

A special memory chip, the **L2 cache**, performs certain tasks more quickly than SDRAM.

AGP Bus provides graphics processors with a dedicated pathway to the main memory of a computer.

PCI Bus is the data highway inside a computer that hard drives, graphics cards and other internal devices send data to each other.

ISA Bus is the data highway inside a computer by which modems, networking cards and other devices send data to each other.

Parallel Port transmits and receives data eight bits at a time, over eight wires and is faster than a serial port. It connects devices such as scanners, printers, external drives, etc. to a computer.

Serial Port transmits and receives data one bit at a time through one wire in a series and is slower than a parallel port. It connects printers, mouses and other devices to a computer.

USB Port (Universal Serial Bus) connects devices such as scanners, printers, mice, keyboards and speakers to a computer.

Power Supply converts the AC voltage from the wall socket to DC voltage that powers the computer circuits.

CMOS battery is what provides the power to a special battery powered memory

— **CMOS** (Complementary Metal-Oxide Semiconductor technology) — that a computer uses to keep track of its particular configuration along with the date and time.

There are several competing "mass" storage technologies that allow users to record as well as read data, including removable hard drives, high-capacity diskettes, cartridges, tape drives, writeable CD's, rewriteable CD's, DVD ROMS, and other numerous devices, all designed to make your life easier.

✪ CPU

Understanding how computers actually work begins with an understanding how the **Central Processing Unit (CPU)** affects the performance of a computer. The CPU (also "microprocessor or "processor") is a vital component that is:

Central — it is the center of computer's data processing.

Processor — it processes (moves and calculates) data.

Unit — it is a chip that contains millions of transistors.

The CPU is the silicon chip (also "silicon wafer"), or processor that makes everything work together. Without the CPU, there would be no Personal Computer (PC). The CPU is the "chip" that performs data manipulation in your computer. A CPU is the brain of the computer. It has millions of switches that control the flow of data; the data is coded into bits by the switches — one bit corresponds to a switch inside the computer that can be "on" or "off". Computer programs called "software" control the switches.

CPUs continually undergo development and for years have doubled their performance about every 18 months. There is no indication that this trend will stop.

CPUs came into being around 1971, when a then unknown company, Intel, combined multiple transistors to form a central processing unit. It was not until years later that the first personal computers came into being. Personal computers are now built using different makes and models of CPUs and although Intel dominates the market, it is certainly not the only company making them. The first CPUs could only work with whole numbers, but once the CPUs reached the Pentium level a mathematical co-processor, called the Floating-Point Unit (FPU), was added to the CPUs architecture for better math processing.

The CPU is situated on the computer's motherboard and its work is primarily calculations and data transport. Before it can be processed, the data you enter into the computer is transmitted along a path to the CPU called the system bus. The system bus feeds streams of data (consisting of the data itself and instructions on how to handle the data) to the CPU. You've probably heard of "program code." That's what the instructions are — program code! Data is what you input via a mouse or a keyboard, i.e., a letter to the telephone company. When you print that letter you actually use "program code" by sending a print instruction to the CPU.

Your computer's CPU is, in all probability, "8086 compatible" which means that your computer's various programs communicate with the CPU in a specific family of instructions, which were written for the Intel 8086 processor. This is also known as "the IBM compatible PC." All IBM compatible processors, no matter how advanced handle the "8086 instruction format."

Many of the older CPUs had a Complex Instruction Set Computer (CISC), which means the computer could understand many complex instructions. However, Reduced Instruction Set Computer (RISC) is used in most newer CPUs. RISC is exactly what it says - the instructions are brief and the same length (for example 32 bit long) processing much faster than the CISC instructions.

You now understand that a CPU is a data processing gadget that is mounted on a printed circuit board called the motherboard. You know that most of the data processing takes place inside the CPU and that the data is transported via the system bus. However, we have not addressed what determines the speed of the CPU. For example, in the Pentium III, "500 MHz" is the clock frequency. What this means is that a small crystal located on the motherboard constantly ticks to the CPU at a steady number of clock ticks per second and at a "tick" something happens in the CPU. Thus, the more ticks per second (frequency) – the more data that is processed per second.

One problem with high clock frequencies is that when the frequency gets too high it becomes an expensive proposition to design the rest of the computer's equipment so it can keep pace. Our manufacturing geniuses came up the solution — split the clock frequency in two, this is called clock doubling. Clock doubling uses a high internal clock frequency for the CPU and a lower external clock frequency for the system bus which is where the CPU exchanges data with the RAM and the Input/Output (I/O) units.

✪ MEMORY

The next important component of a computer is to be found in the chips where data is stored, i.e. memory. There are many kinds of memory necessary for the operation of a computer, such as **ROM** (read only memory) PROM (programmable read only memory), Flash Memory, which is a type of PROM and cache Memory. The only memory you will need to make any decision about is the **RAM** and cache. The CPU can access data in any of the RAM cell locations in any order and it writes to the memory in any order. If you add additional RAM to a computer, you can increase its ability to process programs and applications at a faster speed. There are several types of RAM, you only need to remember one — SDRAM or whatever is the most recent incarnation (forget the others when considering server configuration).

Just remember, when dealing with memory — more is better. In discussing RAM you need to know that the size of memory is measured in bytes and each byte has its own address. For example, 1 byte is 8 bits (remember the switches in the CPU is a bit), so a kilobyte ("KB") is 1,024 bytes, megabyte ("MB") is 1,048,576 bytes, gigabyte ("GB") is about a billion bytes and there is even a terabyte which is about one trillion bytes.

✪ CACHE

CPUs run much faster than everything else in the computer, which means that a computer is designed to ensure that the processor is not slowed down by the devices it works with. Slowdowns mean wasted processor cycles, where the CPU can't do anything because it is sitting and waiting for data. There is a special kind of fast RAM, like SRAM, that a CPU needs; it is usually referred to as the cache. The cache holds the data that is likely to be needed next by the CPU. The cache operates as a buffer between the CPU that is very fast and the system RAM (memory), which is slow – it's not really slow, the CPU is just very fast.

There are different types of cache in a computer, all acting as a buffer for often-used data to enhance the computer's performance. Each layer is closer to the CPU and faster than the layer below it and each layer caches the layers below it, due to its increased speed relative to the lower levels. By utilizing this cache system, when the CPU needs something from memory, it gets it as soon as possible. Here is how it works: The CPU gets a data request and it goes to the L1 cache because it's the fastest. If it finds the data there it uses it and there is no performance delay. If not, the CPU goes to the L2 cache and if it finds it there the CPU goes on with little delay. If the CPU doesn't find the needed data, it sends a "read request" to the system RAM where the data may be stored or the system RAM may have to go to a disk (Hard, floppy, CD-ROM) to retrieve the data thereby slowing the CPU's performance.

Level 1 (L1) Cache (sometimes referred to as internal cache) consists of the high-speed memory that is built into most CPUs. It is small, generally from 8 KB to 64 KB, but fast, it runs at the same speed as the CPU. By using L1 cache, the CPU can access often-used data more quickly. The amount of L1 varies but is not upgradeable.

Level 2 (L2) Cache, also called the "burst" or the "pipeline" cache; and when "cache" is referred to without qualifiers or as "system cache" or "external cache", it means the Level 2 cache that is placed between the processor and the system RAM. The L2 is usually separate from the CPU and situated on the Motherboard. It is larger and a little slower than the L1 Cache, the size varies (usually between 512 KB to 2 MB) and, unlike the L1, it is usually upgradeable. The L2 works in conjunction with the CPU's internal cache (L1) to provide maximum performance. This cache is where oft-used data is retained in memory as a way to help the CPU so that it seems to run even faster when this oft-used data is accessed. When considering the L2, remember that the rule of thumb is "the bigger the better."

Disk cache is the part of the system RAM used to cache reads and writes to a drive (hard or external). It is not the size of disk cache that is important, but the organization of the cache itself ("write/read cache" or "look ahead cache"). Disk cache has the slowest speed since system RAM is slightly slower than the L1 or L2 and the drives themselves are much slower than the system RAM.

✪ HARD DRIVES

Your next major consideration is the hard drive. Your hard drive is the computer's storage area — the filing cabinet. All files (operating system, applications and data) are stored on the hard drive. Hard drives are measured in "Gigabytes". The bigger your hard drive is, the more files you can load. The hard drive has a tremendous impact on the computer's overall performance causing delay while data is pulled off the drive. For example, drive access time is measured in milliseconds and RAM in nanoseconds.

The **EIDE** Enhanced Integrated Drive Electronics) drives allows fast transfers and large capacities. The computer's system RAM is used for storing the drive's firmware (software or BIOS). When the drive powers up, it reads the firmware.

The **SCSI** (Small Computer Serial Interface) drive is the fastest with the largest capacity and the highest transfer rate available on the market. This is not to say that the EIDE drive isn't capable of the same, it's just that high-end drives with high capacity and high performance are built for servers. The power of SCSI is that several devices can use its interface simultaneously and the multitasking environment of servers is ideal for SCSI since there is frequent simultaneous access.

✪ THE INTERFACE

One of the factors that affect the speed of a hard drive is the interface. Currently there are two common interfaces: EIDE and SCSI. The EIDE controller is integrated with

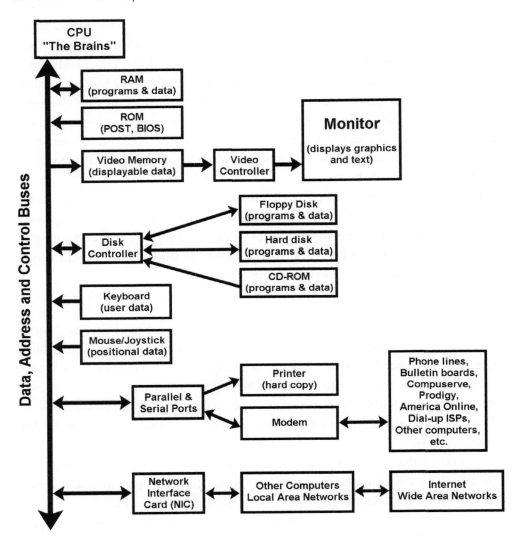

Figure 29. Schematic diagram of the infrastructure of the typical computer.

the motherboard. The SCSI requires an extra controller since most motherboards don't have integrated SCSI controllers.

The EIDE interface can connect a total of four devices to a computer bus such as 2 hard drives, a CD-ROM and a DVD.

The SCSI interface can connect up to 15 devices to a computer bus. There are several types of SCSI interfaces — 8-bit, 16-bit and even 32-bit. The transfer rate started at 5 Mbps for the old SCSI 1 evolving in increments up to the 16-bit, 40 Mbps Ultra2 SCSI, then the 16-bit, 80 Mbps Ultra2 Wide SCSI. The latest incarnation is the 32-bit, 160 Mbps Ultra3 SCSI or Ultra1 60. For normal computer operation, the performance of a drive receives only a small boost from the SCSI interface but for a website, you should seriously consider the SCSI.

Glossary

A

ACLS (Access Control List Service) – Restricts access to computer resources such as files and directories.

Active Server Page (ASP) – A dynamically created web page that employs ActiveX scripting. When a surfer requests an ASP page through their browser, the web server generates, in real-time, a page with HTML code, which it then sends to the browser.

Active X – An architecture that lets a program (the Active X control) interact with other programs over a network such as the Internet.

Active X Controls – The interactive objects in a web page that provides interactive and user-controllable functions.

American Registry for Internet Numbers (ARIN) – One of three Regional Internet Registries (RIRs) worldwide that collectively provide IP registration services to all regions around the globe. The other two RIRs are APNIC (Asia Pacific network Information Center) and RIPE (Reseaux IP Europeens).

API – See Application Programming Interface.

Applet – Small software programs that can be downloaded quickly and used by any computer equipped with a Java-capable browser. See also, Java.

Application – A software program that does some type of task. MSWord, Netscape, Winzip, and anti-virus programs are a few examples of an application.

Application Programming Interface (API) – Software that an application uses to request and carry out lower-level services performed by a computer's operation system. In short, an API is a hook into software. An API is a set of standard software interrupts, calls and data formats that applications use to initiate contact with network services, mainframe communication programs, etc. Applications use APIs to call services that transport data across a network.

Application Service Provider (ASP) – A third party that manages and distributes software-based services and solutions to its customers from a centralized server base.

Architecture – Refers to the overall organizational structure of a given system, i.e. processor architecture or proprietary architecture. Central to the architecture is the decision about the selection of structural elements and their behavior, as defined by collaborations of larger subsystems, therefore the architectural style is the definitive guide for the system.

Autoresponders – A mail utility that automatically sends a reply to an e-mail message. They are used to send back boilerplate information on a topic without having the requester do anything more than e-mail a particular address. They are also used to send a confirmation that the message has been received.

ASP – See Active Server Page and Application Service Provider.o

B

Backbone – See Internet Backbone.

Bandwidth – As used in this book, the term "bandwidth" refers to how much information can be carried in a given time period (usually a second) over a communications link.

Bit – The smallest unit of information a computer can process and the basic unit in data communications. Bits compose a byte.

Blog – Also known as a "weblog, at its most basic, a blog is a journal that is available on the Web. The writer (commonly known as a "blogger") typically updates his or her online journal using software that enables even people with little or no technical background to update and maintain their blog. Postings on a blog are usually arranged in chronological order with the most recent additions viewed first.

Blogger – Someone who keeps a blog.

Blogging – The activity of updating a blog.

Bookmark - Saving a web location (typically via a feature offered by a web browser) in order to be able to go back to it at a later point with a simple click. The term, "bookmark," is used because the action is similar to sticking a paper bookmark in a book. Most web browsers provide the user with the capability to "save" a web location (i.e. address) so the user can quickly access the web address when it is needed.

Bot – See Spider.

Brick-and-mortar - A traditional business (usually retail) with actual buildings, manufacturing plants, customer service centers, or distribution facilities.

Bus – All computers use buses — a collection of wires through which data is transmitted from one part of a computer to another. There are two common buses inside a PC – the older ISA bus, capable of transmitting only five megabytes per second and the newer PCI bus, capable of transmitting up to 132 megabytes per second.

Business Rules – A conceptual description of an organization's policies and practices enabling them to automate their polices and practices, and to increase consistency and timeliness of their business processing.

Byte – A set of bits of a specific length that represent a value, in a computer coding scheme. A byte is to a bit what a word is to a character.

C

C – A very powerful programming language that operates under Unix, MS-DOS, Windows (all flavors) and other operating systems.

Cache – To store data on a disk for quick and easy retrieval instead of retrieving it each time it is requested. In order to conserve bandwidth, large ISPs and businesses will cache popular web pages.

Call center – A functional area that is operated by a company to service is customers/clients, or it can be an outsourced, separate facility. Either way, a call center exists solely to answer inbound or place outbound telephone calls.

ccTLDs – Short for country-code top-level domain, a domain name suffix that is linked to a specific nation. It is comprised of two letters such as .uk for the United Kingdom.

Cells – The smallest component of a table. In a table, a row contains one or more cells.

CGI (Common Gateway Interface) – A predefined way in which CGI programs or scripts communicate with a web server. See also Server-side language.

CGI bin – A directory on a web server in which CGI programs (scripts) are stored.

CGI Script – A program consisting of small but highly potent bits of computer code that is usually executed on a web server so as to provide interactivity to web pages.

Chargen (CHARacter GENerator) – This term refers to an utility that provides an approximate speed for a computer's Internet connection in characters per second (with compression taken into account).

Chat – A term that is used to describe a type of computer-based communication. When you "chat" with someone you usually type your communications into a computer as text, although voice and video images can also be used. One person sends a message and another person or a group of people can read it and respond to it. This happens in real-time so, unlike e-mail, there is little time delay before receiving a response to a message.

CISC (Complex Instruction Set Computer) – A microprocessor architecture that favors robustness of the instruction set over the speed with which individual instructions are executed.

Client/Server – The client is a PC or program "served" by another networked computing device in an integrated network, which provides a single system image. The server can be one or more computers with numerous storage devices.

Cluster – A group of computers and storage devices that function as a single system sharing one or more panel runs and working in a fault-resilient manner, allowing increased effectiveness and efficiency of security, administration and performance.

Comment Tag – Used to insert comments in an HTML document. Comment tags are ignored by browsers (example: <!– text –> or <Comment>text</comment>).

Connectivity – The property of a network that allows dissimilar devices to communicate with each other. It also refers to a program's or device's ability to link with other programs and devices.

Content – A website's offerings — products, graphics, marketing materials, banner adds — anything that is contained within the pages of a website.

Cookie – An HTTP header that contains a string that a browser stores in a small text file in the Windows/Cookies directory (for Microsoft Internet Explorer) or in the Users folder (for Netscape Navigator) on a computer's hard drive. Cookies store information supplied by a user to be accessed at a later period in time but it is important to state that a cookie can't interact with other data on a computer's hard drive.

CPE (Customer Premises Equipment or Customer Provided Equipment) – Refers to equipment on the customer's premises, which had been bought from a vendor.

CPU (Central Processing Unit) – This term refers to a programmable device that can process digital information.

Crawler – See Spider.

Cross-selling – Offering a product similar to the one the customer is interested in if the chosen product is unavailable.

CSU/DSU – See DSU/CSU.

CTR (Click-Through-Rate) – The ratio of impressions to click-throughs.

D

Database – Software that enables the storage of data, the retrieval of that data, and the ability to add to and change the stored data when necessary.

Database Publishing – Allows businesses to leverage existing data and data management assets. Many of today's database applications can create files usable by electronic publishing software. By establishing communication the database can continue managing data, and the publishing system can be used as an information synthesis tool to gather data from a variety of sources (databases, graphics, and text) and present it in a single, cohesive document.

DBMS (DataBase Management System) – Software that controls the organization, storage, retrieval, security and integrity of data in a database. It accepts requests from the application and instructs the operating system to transfer the appropriate data.

Digital Certificate – A small piece of unique data used by encryption and authentication software. A digital-based ID that contains a user's information. It accomplishes this by attaching a small file containing the certificate owner's name, the name of its issuer and a public encryption key to the information that is transmitted over the Internet.

Directories – A directory is basically a manual entry database system for which a URL is submitted along with a descriptive title and summary for the website.

Disk Pack – An assembly of magnetic disks that can be removed from a disk drive along with the container from which the assembly must be separated when operating.

Domain name – Unique address of a website. The address that gets you to a website, and consists of a hierarchical sequence of names separated by dots (periods). Also known as a web address. It can identify one or more IP addresses. See URL.

Domain Name System (DNS) – People use the domain name addressing system, whereas computers use the Domain Name System. A DNS takes a domain name address and automatically translates that address into a numerical IP address via DNS servers (also called name servers), maintain databases that contain the web addresses/domain name. The IP address is then used by services such as packet-routing software and computers to get data packets to their destination. Note: DNS is also the acronym for the "Domain Name Service," which is the Internet utility that implements the Domain Name System.

Doorway page – Also known as bridge page, portal page, jump page, gateway page, entry page, a doorway page is a web page that is designed primarily for search engines, not for human beings. The term can also refer to a web page that serves only as an entry point through which users click to go to the actual homepage. In the latter instance, a doorway page is created for each of the leading search engines. The only time a user sees one of these doorway pages is when they click on a link from a search engine's result page.

Dot-com – A company with operations are entirely or primarily internet-based. While a dot-com may deliver all their services over an Internet interface, the company's products may be delivered through traditional channels.

DSL (Digital Subscriber Line) – A technology that delivers high-bandwidth over ordinary copper telephone lines.

DSU/CSU (Digital or Data Service Unit/ Channel Service Unit) – Communication devices that connect an in-house line to an external digital circuit, such as a T1, DDS, etc. A DSU/CSU is similar to a modem, although it connects a digital circuit rather than an analog one.

Dynamic Web page – The dynamic change in the contents of a web page through the use of a separate file wherein the current contents of that file is displayed on all pages connected to the underlying database whenever a browser requests a web page.

E

E-1 – See T-1.

ECML (Electric Commerce Modeling Language) – A universal format for online checkout form data fields. ECML provides a simple set of guidelines that automate the exchange of information between consumers and web-based merchants.

Echo – A command in a software program that sends data to another computer which "echoes" it back to the user's screen display, allowing the user to visually check if the other computer received the data accurately.

E-Business – The electronic exchange of information including e-commerce.

E-Commerce – Buying and/or selling electronically over a telecommunications system. In doing so every facet of the business process is transformed: Pre-sales, updating the catalog and prices, billing and payment processing, supplier and inventory management, and shipment. By using e-commerce a business is able to rapidly process orders, produce and deliver a product/service at a competitive price and at the same time minimize costs.

EDI – See Electronic Data Interchange.

EIDE (Enhanced Integrated Drive Electronics) – A hard drive that allows fast transfers and large storage capacities. The computer's system RAM is used for storing the drive's firmware (software or BIOS). When the drive powers up, it reads the firmware.

EJB (Enterprise JavaBeans) – A Java API developed by Sun Microsystems that defines a component architecture for multi-tier client/server systems. See also Java.

Electronic Commerce Modeling Language (ECML) – Technology that provides a set of hierarchical payment oriented data structures that enable automated software, including electronic wallets (e-wallets) from multiple vendors, to supply needed data in a uniform manner.

Electronic Data Interchange (EDI) – A series of standards that provide computer-to-computer exchange of business documents between different companies' computers over the Internet (and phone lines). EDI allows for the transmission of purchase orders, shipping documents, invoices, invoice payments, etc. between a web-based business and its trading partners. EDI standards are supported by virtually every computer and packet-switched data communications company.

E-mail Management System (EMS) – Intelligent automatic customer e-mail management software used to automate the management and dispatch of emails. An email management system, for example, can send an acknowledgement of a received email automatically, automatically route an email to a specific mailbox or mailboxes, provide ways to improve email interactions with customers, etc.

Encryption – A system of using encoding algorithms to construct an overall mechanism for sharing sensitive data. Encryption is commonly used for the translation of data into a secret code.

ERP (Enterprise Resource Planning) – A business management system that integrates all aspects of a business, such as, product planning, manufacturing, purchasing, inventory, sales, and marketing. ERP is generally supported by multi-module application software that helps to manage the system and interact with suppliers, customer service, and shippers, etc.

E-wallet – The electronic equivalent of a wallet for e-commerce transactions. A digital wallet (e-wallet) can hold digital money that is purchased similar to travelers' checks or to a prepaid account like an EZPass system, or it can contain credit card information. The wallet may reside in the user's machine or on the servers of a web payment service. When stored in the client machine, the wallet may use a digital certificate that identifies the authorized card holder.

Extranet – A term that refers to a private, TCP/IP-based network that allows qualified users from the outside to access an internal network.

F

FAQ (Frequently Asked Questions) – A term that refers to an online document that poses a series of common questions and answers on a specific topic.

Ferroresonant transformer – A transformer that regulates the output voltage by the principle of ferroresonance. This occurs when an iron-core inductor is part of an LC circuit and it is driven into saturation, causing its inductive reactance to increase to equal the capacitive reactance of the circuit.

Finger – A standard protocol. A program implementing this protocol lists who is currently logged in on another host. It is a computer command that displays information about people using a particular computer, such as their names and their identification numbers. (Integrated finger is a common Unix network function that reports information relating to a user after entering his or her e-mail address.)

Firewall – Hardware and/or software that sit between two networks, such as an internal network and an Internet service provider. It protects the network by refusing access by unauthorized users. It can even block messages to specific recipients outside the network.

Firmware – Software which is constantly called upon by a computer so it is stored in semi-permanent memory called PROM (Programmable Read Only Memory) or EPROM (Electrical PROM) where it cannot be "forgotten" when the power is shut off. It is used in conjunction with hardware and software and shares the characteristics of both.

FPU (Floating Point Unit) – A formal term for the math coprocessors found in many computers. The modern computer has the FPU integrated with the CPU.

Frames – A programming device that divides web pages into multiple, scrollable regions; this is done by building each section of a web page as a separate HTML file and having one master HTML file identify all of the sections.

Frame Relay – A packet-switching protocol for connecting devices on a Wide Area Network (WAN). Frame Relay supports data transfer rates at T-1 (1.544 Mbps) and T-3 (45 Mbps) speeds.

Framesets – See Frames.

F-Secure SSH – Provides for secure UNIX shell logins. SSH creates encrypted connections that protect confidential information, such as passwords, from exposure to network eavesdroppers.

FSP – See Fulfillment Service Provider.

FTP (File Transfer Protocol and File Transfer Program) – Allows users to quickly transfer files to and from a distant or local computer, list directors, delete and rename files on the distant computer. FTP the program is a MS-DOS program that enables transfers over the Internet between two computers.

Fulfillment Service Provider (FSP) – A third-party business that outsources its warehousing and distribution services and expertise to other companies such as e-commerce websites.

G

Gateway – An electronic repeater device that intercepts and steers electrical signals from one network to another. In data networks, gateways are typically a node that connects two otherwise incompatible networks and often perform code and protocol conversion processes.

GIF (Graphics Interface Format) – A format for encoding images into bits so a computer can read the file and display the image on a computer screen.

gTLD (generic Top Level Domain) – A small set of top-level domains that do not carry a national identifier, but denote the intended function of that portion of the domain space. For example, .com was established for commercial users and .org for not-for-profit organizations.

GUI (Graphical User Interface) pronounced "gooey" – A program with a graphical interface that can take advantage of a computer's graphics capabilities thereby making the program easier to use.

H

Hacker – An unauthorized person who breaks into a computer system to steal or corrupt data.

Hardware – Objects that go with the computing environment that can be touched. For example, modems, interface cards, floppy disks, hard drives, monitors, keyboards, printers, motherboards, memory chips, etc.

Home page – The main page of a website, usually serving as an index or table of contents to other documents stored on the web server.

HTML (HyperText Markup Language) –

Used to create documents on the World Wide Web by defining the structure and layout of a web document through the use of tags and attributes thereby determining how documents are formatted and displayed.

HTTP (HyperText Transfer Protocol) – Defines how messages are formatted and transmitted over the World Wide Web, and what actions web servers and browsers should take in response to various commands.

Hypertext – A type of system in which objects, whether they are text, graphic files, sound files, programs, etc., can be creatively linked to each other.

I

IDSN (Integrated Services Digital Network) – A set of communications standards that enable a single copper wire or optical fibre to carry voice, digital network services and video.

Image Map or Imagemap – Clickable images. The image is a normal web image (usually in GIF or JPEG format). The map data set is a description of the mapped regions within the image. The host entry is HTML code that positions the image within the web page and designates the image as having map functionality.

Infrastructure – The interconnecting hardware and software that supports the flow and processing of information.

Input/Output (I/O) Unit – Any operation, program, or device that transfers data to or from a computer, such as disks (floppy, hard, or writable CD-ROMs, etc.). I/O units can also consist of single function operations such as Input-only devices such as keyboards and mouses and output-only devices such as printers.

Intranet – An internal TCP/IP-based network behind a firewall that allows only users within a specific enterprise to access it.

Internet – A public global network of computers that exchange data.

Internet Backbone – The worldwide structure of cables, routers and gateways that form a super-fast network. It is provided by number of ISPs that use high-speed connections (T-3s, Ocs) linked at specific interconnection points (national access points).

Internet Address – A registered IP address assigned by the InterNIC Registration Service.

IP Address – A unique identification consisting of a series of four numbers between 0 and 255, with each number separated by a period, for a computer or network device on a TCP/IP network.

J

Java – A high-level object-oriented programming language similar to C++ from Sun Microsystems designed primarily for writing software to leave on websites which is often downloadable over the Internet. Java is basically a new virtual machine and interpretive dynamic language and environment.

Java Applet – Small Java applications that can be downloaded from a web server and run on a user's PC by a Java-compatible web browser.

Java Script – An open source scripting language developed by Netscape (independent of Sun's Java) that enables interactive websites by interacting with HTML source code.

JDBC (Java DataBase Connectivity) – A Java API that enables Java programs to execute SQL statements similar to ODBC.

JPEG (Joint Photographic Experts Group) also JPG – A compression technique used in editing still images, color faxes, desktop publishing, graphic arts and medical imaging. Although it can reduce image files to approximately 5% of their normal size, some detail is lost in the compression.

Just-in-Time Inventory Management – An inventory management system whereby products arrive at the business' warehouse just in time to ship.

K

Kerberos – A security system that authenticates users but doesn't provide authorization to services or databases, although it does establish identity at log-on.

Keyword – In database management, a keyword is an index entry that identifies a specific record or document. In programming, keywords (sometimes called reserved names) can be commands or parameters, which are reserved by a program because they have special meaning. On the World Wide Web keywords are the terms that you enter into the search field of a search engine or directory.

Knowledge base – As used in this book, a knowledge base is a database that contains common questions and the answers to those questions. Although a knowledge base sounds a lot like a FAQ section on a website, its a bit more than that – a knowledge base usually is searchable, and often offers special options to make it more useful to users, such as natural language searching, options to add searchable keywords that don't appear in the question or answer (but which apply), etc. The Microsoft website's technical section is built upon knowledge base technology.

L

LAN (Local Area Network) – A short distance data communications network consisting of both hardware and software and typically residing inside one building or between buildings adjacent each other – thus allowing all networked devices to share each other's resources.

Link – On the World Wide Web, a link is a reference to another website or web page or document and it takes you to the other website, web page or document when you click on it.

M

Macros – Small simple programs written to automate specific tasks are often referred to as "macros."

Mail List – A program that allows a discussion group based on the e-mail system.

MAN (Metropolitan Area Network) – Two or more LANs links together so resources between the LANs can be shared.

Merchant Account – A business account at a financial institution that functions as a clearing account for credit card transactions.

Meta tags – HTML code between the <HEAD> and </HEAD> section of a web page. Meta tags are accessed by a search engine's spider and used by search engines to describe your entire website and individual web pages. Meta tags are either in the form of keywords or a descriptive phrase.

Micropayments – A business transaction type, which specializes in the sub-dollar range. Although each transaction is a small amount, it can add up to a sizable market because of global access of the Internet.

Mirroring – As used in this book, the term

"mirroring" refers to the creation of multiple websites that are exact duplicates of an existing website.

Motherboard – A circuit board that everything (adapter cards, CPU, RAM, etc.) plugs into and therefore provides the actual physical connection between the different components that make up the computer.

N

Navigation – Traveling from place to place on a web page, a website or the Web, from information to information. It can also mean to search for information from a menu hierarchy or hypertext. Navigation in a hypertext environment (the Web) is a physical experience of scrolling, scanning and clicking, moving over the text with your mouse pointer and actively clicking on hyperlinks.

Netstat – A utility that provides statistics on the network components, i.e., it shows the network status. It displays the contents of various network-related data structures in various formats. For example, it displays: all connections and listening ports (server connections are normally not shown), ethernet statistics, addresses and port numbers in numerical form (rather than attempting name look-ups), per-protocol statistics (shows connections for the protocol specified by protocol), and the contents of the routing table.

O

ODBC (Open DataBase Connectivity) – A standard database access method developed by Microsoft Corporation that allows databases such as dBASE, Microsoft Access, Fox-Pro and Oracle to be accessed by a common interface independent of the database file format.

ODBMS – Object-oriented Database Management System.

Object-oriented Database Management System – A database management system (a program that lets one or more users simultaneously create and access data in a database) that supports the modeling and creation of data as an object.

Operating system – The software that controls all of a computer's operations including memory allocation, accessing disk drives and calling applications is known as the operating system.

P

Packet – A bundle that contains data and certain control information including a destination address which is transmitted on a network.

PC (Personal Computer) – A computer designed for use by a single person versus simultaneous use by more than one person.

PDF (Portable Document Format) – Adobe technology that provide a popular method for formatting documents in such a way that they can be viewed and printed on multiple platforms using the freely available Adobe Acrobat reader.

PERL (Practical Extraction and Report Language) – An interpreted scripting programming language, commonly used in writing CGI scripts because it is a superb text manipulator.

Ping (Packet InterNet Groper)– A program used to test whether a particular network destination on the Internet is online by repeatedly bouncing a "signal" off a specified address and seeing how long that signal takes to complete the round trip. A common Unix network function that reports on whether another computer is currently up and running on the Internet as well as how long the ping takes to reach the computer.

Plug-ins – Software modules that run on the viewers' local machine and add to the functionality of an application, such as a browser.

Point of Presence – An access point to the Internet that includes includes routers, digital/analog call aggregators, servers, and frame relays or ATM switches.

POP – See Point of Presence.

Pop-up Window – A second browser window that "pops up" when called by a link, a button or an action.

Processor Architecture – A term used to refer to the over all organizational structure of a computer's processor. The main elements of any processor architecture are the selection and behavior of the structural elements and the selected collaborations that form larger subsystems that guide the workings of the entire processor.

Protocol – A set of rules governing the format of messages that are exchanged between computers and people.

Proxy Server – A server that rests between the client and the server to monitor and filter the traffic traveling between them. It can boost web browser response time by storing copies of frequently accessed web pages.

Q

Query – A process by which a web client (e.g. a web browser) requests specific information (based on a character string) from a web server. Usually a query is simply a database search for a particular keyword or phrase.

R

Rack Unit – A vertical shelving system to mount servers.

RAID (Redundant Array of Independent (or Inexpensive) Disks) – A system designed to link the capacity of two or more hard drives that are then viewed as a single large virtual drive by the RAID management system. This allows for improved data storage reliability and fault tolerance.

RAM (Random Access Memory) – Computer data storage that comes in the form of chips that can be accessed randomly, i.e., any byte of a chip can be accessed without touching the preceding bytes. RAM is the most common memory found in computers, printers, etc.

RDBMS (Relational DataBase Management System) – A program that enables one or more people to simultaneously create, update and administer a relational database.

Real-Time – Occurring immediately (as opposed to simultaneously as in real time). The data is processed the moment it enters a computer, as opposed to BATCH processing, where the information enters the system, is stored and is operated on at a later time.

Request for Proposal (RFP) – A formal request to a third-party outsourcer describing everything you want the outsourcer to accomplie and requesting the outsourcer to write a proposal outlining how he or she would go about meeting your demands and the costs thereof.

RFP – See Request for Proposal.

RISC (Reduced Instruction Set Computer) – A microprocessor that is designed to favor the speed at which individual instructions execute over the robustness of the instructions; i.e., it performs a smaller simpler set of operating commands so that the computer can operate at a higher speed.

Router – A high intelligent device that connects like and unlike LANs (Local Area Networks), MANs (Metropolitan Area Networks) and WANS (Wide Area Networks), which are protocol sensitive and which can forward packets between the connected networks. As software, it is a system level function that directs a call to an application.

RSS – An acronym that stands for either "Really Simple Syndication" or "RFD (Resource Description Framework) Site Summary." RSS is a protocol that provides an open method of syndicating and aggregating web content. It is the code that underlies a weblog.

S

SCSI (Small Computer System Interface) pronounced "scuzzy" – A standard for a bus (with its own controller and microprocessor) and interface that allows faster communication between the input/output devices and the computer's main processor and daisy chaining of up to 7 different devices.

Script – A type of computer code that can be directly executed by a program that understands the language in which the script is written.

Search engine – A database system designed to index Internet addresses via a schema that allows submission of a URL and through a defined process the search engine includes the submitted URL into its index.

Secure Electronic Transactions (SET) — A technology that is used to encrypt credit card numbers so that only designated banks and credit card companies can read the information. SET requires the website operator to obtain a special certificate from his or her bank, and then the customers must install special software on their computer to take advantage of SET protection.

Secure Socket Layer (SSL) Server – A server that utilizes the SSL protocol in order to encrypt transmission over TCP/IP networks such as the Internet.

Server – A computer that manages network resources. For example, a web server has a very fast permanent connection to the Internet and subsystems to protect against power outages, hackers and system crashes. A database server manages and processes the database and database queries.

Server cabinet – A metal cabinet designed to house rack-mounted servers (some also house tower configured systems). The cabinet will usually have a slotted front door, perforated steel rear door and top panel, with room for fans or blowers, and lift off side panels.

Server-side language – Also known as CGI script, server-side languages are programming languages (e.g. Perl, ColdFusion, ASP, PHP, etc.) that are designed to run on a server rather than a client such as a web browser.

SET (Secured Electronic Transactions) – A standard that enables secure credit card transactions on the Internet thereby making the theft of credit card numbers via the Internet much more difficult.

Shell account – A login account on a UNIX server.

Site description – A description of how a website functions from web page to web page or section to section. A site description works in conjunction with the storyboard to provide a detailed explanation of workflow, data flow, and other items that may not be readily apparent in the storyboard.

SMTP (Simple Mail Transfer Protocol) – This refers to a TCP/IP protocol that is commonly used for sending e-mail between servers. The majority of the e-mail systems in use today send mail use SMTP to send mail via the Internet.

SNMP (Simple Network Management Protocol) – This term refers to a specific network monitoring and control protocol. Data is passed from SNMP agents that are hardware and/or software processes reporting activity in a network device such as a hub, router, bridge, etc. to the computer administering the network. The SNMP agents return information contained in a MIB (Management Information Base), which is a data structure that defines what is obtainable from the device and what can be controlled, i.e., turned off, on, etc.

Software – Computer instructions or data – anything that can be stored electronically is software.

SPARC – Sun Microsystems' open RISC-based architecture for microprocessors. SPARC is the basis for Sun's own computer platforms and it's licensed to third parties.

Spider – A special program (also referred to as "bot" or "crawler") utilized by search engines to index websites. There are two spider classes — deep and shallow. The deep spider drills through the entire website and then either finds the URL and copies the file or finds a single directory within the URL and copies the file. A shallow spider can do one of two things, it can either spider the URL

given and stop, or only spider those URLs it finds within a single level of directories.

SQL (Structured Query Language) pronounced "sequel" – A database language used for creating, maintaining and viewing database data.

SSL (Secure Socket Layer) – A transport level technology developed by Netscape for authentication and data encryption between a web server and a web browser.

SSL Server Certificate – Also known as a Digital Certificate, it is a small piece of unique data used by encryption and authentication software that enables SSL encryption on a web server. This allows a website to accept credit card orders securely.

Stickiness – Website elements that keep visitors not only within a website's pages, but also keeps them coming back for more. Stickiness is the byproduct of a website that is designed around both good content and good design elements.

Storyboard – The pictorial representation of the screen elements and their operations for every web page, which taken as a whole, constitutes a website.

Systat – A program owned by SPSS that provides powerful statistical techniques through its convenience in handling data, selecting and defining procedures to use, and formatting output.

T

T-1 – A high-speed digital connection capable of transmitting data at a rate of approximately 1.5 million bps (bits per second). In Europe the same basic technology is known as "E-1."

Tables – A collection of data arranged in

rows and columns and in which each item is arranged in relation to the other items.

Targeted text link – Text you can click on that will transport you to a specific section of a website.

Tcl Scripting Language (pronounced "tickle") – An open-source tool command scripting language that can be embedded within existing C++ code. As such, Tcl is used in the development of many websites.

TCP (Transmission Control Protocol) – A transport layer, connection-oriented, end-to-end protocol that provides reliable, sequenced and unduplicated delivery of bytes to a remote or local user.

TCP/IP (Transmission Control Protocol/Internet Protocol) – A networking protocol (the Internet's protocol) that provides communication across interconnected networks, between computers with diverse hardware architectures and various operating systems.

Telnet – A terminal-remote host protocol. A program that lets one computer connect to another computer on the Internet.

Text-only Browser – A browser that cannot handle hypermedia files. For example, Lynx is a text-only browser that lets you travel from one link on the Web to the next, in sequential order. Lynx gives access to all of the information that the graphical browsers can, just without the pictures or sounds. Netscape and Internet Explorer are graphical browsers that let you see pictures and hear sound.

TLD (Top Level Domain) – Domains of which .com, .net and .org are the most common.

Traceroute – Software to help you analyze what's happening on an Internet connection by showing the full connection path between one website and another Internet address. A common Unix network function that reports the number of hops, or intermediate routers, between a computer and a remote server.

Traffic Allowance – Refers to how many bytes can be transferred from a website per month, i.e., number of megabytes sent to a website's visitors' browsers.

U

UPS (Uninterruptable Power Supply) – Generally a device that allows a system to maintain operation when changes to the power supply would otherwise interrupt the function of that system. They can range from a 9 Volt battery backup to a generator.

Up-selling – Offering customers additional recommendations when they are placing an order.

Uptime – The percentage of time a service (web-hosting service, operating system, etc.) is up and running.

URL (Uniform Resource Locator) – The global address of resources on the Internet. The first part of the address indicates what protocol to use, and the second part specifies the IP address or the domain name where the resource is located.

V

Voice-over-IP (VoIP) – A term that refers to a system for transmitting telephone calls over data networks, including the Internet. VoIP technology provides the ability to carry packetized voice over an IP-based network (e.g. the Internet) with POTS-like (Plain Old Telephone Service) functionality, reliability and voice quality.

VPN (Virtual Private Network) – A secure, encrypted connection between two points across the Internet. It can act as an intranet or extranet, but uses the Internet as the networking connection. Most VPNs are built and run by Internet service providers.

W

WAN (Wide Area Network) – A network that is geographically scattered with a broader structure than a LAN. It can be privately owned or leased, but the term usually implies public networks.

Web – A subset of the Internet that in today's world is accessed via a web browser.

Web address – See Domain Name.

Web hosting service – A third party that leases space on its web servers and use of its other hardware such as UPS, backup, its technical staff, etc., so the lessee's website can be accessed over the Internet.

Weblog – See Blog.

Web page – A document on the World Wide Web that has its own unique URL (Uniform Resource Locator). Websites consist of any number of web pages.

Web page editor – A plain text editor, such as Notepad for Windows offers a place to type in your HTML code so that you can post the file on your website. A more complicated editor can do just about everything for you (so there is no need to know HTML) just drag and drop text, images, etc. onto your page, and the editor writes the code.

Web Server – A computer with data and specific software to operate a website.

Website – Data residing on a computer, which has software running on it that allows the download and presentation of the data to another computer that is permanently connected to the Internet.

Whois – A common Unix network function that queries databases for information about domain names, IP address assignments, and individual names.

WYSIWYG (What You See Is What You Get) pronounced wiz-e-wig – An application that enables you to see on the computer monitor exactly what will appear when the document is printed.

XYZ

XML (eXtensible Markup Language) – A system for organizing and tagging elements of a document specifically designed for web documents. It enables designers to create their own customized tags to provide functionality not available with HTML. XML also has the ability to enable the structured exchange of data between computers attached to the Web, thus allowing one web server to talk to another web server. This means manufacturers and merchants can begin to quickly swap data, such as pricing, stock-keeping numbers, transaction terms and product descriptions.

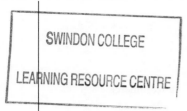

Index

#0310 - 220916 - C0 - 229/179/21 - PB - 9781578203123